W9-AOJ-408

DATE DUE

THE SEARCH
FOR
E. T. BELL

also known as John Taine

SPECTRUM SERIES

The Spectrum Series of the Mathematical Association of America was so named
to reflect its purpose: to publish a broad range of books including biographies,
accessible expositions of old or new mathematical ideas, reprints and revisions
of excellent out-of-print books, popular works, and other monographs of high
interest that will appeal to a broad range of readers, including students and
teachers of mathematics, mathematical amateurs, and researchers.

From Zero to Infinity, by Constance Reid
I Want to be a Mathematician, by Paul R. Halmos
Journey into Geometries, by Marta Sved
The Last Problem, by E. T. Bell (revised and updated by Underwood Dudley)
Lure of the Integers, by Joe Roberts
Mathematical Carnival, by Martin Gardner
Mathematical Circus, by Martin Gardner
Mathematical Cranks, by Underwood Dudley
Mathematical Magic Show, by Martin Gardner
Mathematics: Queen and Servant of Science, by E. T. Bell
Numerical Methods that Work, by Forman Acton
Polyominoes, by George Martin
The Search for E. T. Bell, also known as John Taine, by Constance Reid
Shaping Space, edited by Marjorie Senechal and George Fleck
Student Research Projects in Calculus, by Marcus Cohen, Edward D. Gaughan,
          Arthur Knoebel, Douglas S. Kurtz, and David Pengelley

Mathematical Association of America
1529 Eighteenth Street, NW
Washington, DC 20036
800-331-1MAA   FAX 202-265-2384

THE SEARCH
FOR
E. T. BELL

also known as John Taine

Constance Reid

Published by

THE MATHEMATICAL ASSOCIATION OF AMERICA

©1993 by
The Mathematical Association of America (Incorporated)
Library of Congress Catalog Card Number 93-78369

ISBN 0-88385-508-9

Printed in the United States of America
Current printing (last digit):
10 9 8 7 6 5 4 3 2 1

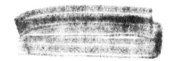

To Dan

Acknowledgments

In writing of my search for E. T. Bell, I have tried to mention in my text individuals and institutions as they provided assistance. Unfortunately it has not always been possible to do so or, in the case of institutions, to cite with sufficient formality. Hence, this special section.

I would like to express here my appreciation for the assistance, not previously noted, of Bill Baum, Gloria Campbell, Carol Champion, Milton Chatton, Steve Clark, Alice Epstein, David Gale, Robert Heilbron, Edwin Hewitt, Andreas M. Hinz, Virginia Kotkin, Paul Langley, Alice Leighton, Gary Lundell, Arthur Nightall, Jerry Savage, Lee Stout, Martha Tucker, and Patricia White. The following members of the science fiction community also responded generously to my queries: Patrick Adkins, Ray Beam, Lloyd Biggle Jr., Thomas Clareson, Donald M. Grant, Hal W. Hall, Charles D. Hornig, Sam Moskowitz, Peter Nicholls, Frederik Pohl, A. Langley Searles, Alfred J. van der Poorten, and Gary K. Wolfe.

Because Bell did not keep copies of his own letters or save those that he received, I am fortunate that the following institutions have given me permission to quote from sizable groups of letters in their possession: The Archives, The California Institute of Technology (correspondence with A. D. Michal and Muriel Rukeyser); The David Eugene Smith Papers, The Rare Book and Manuscript Library, Columbia University; The William Marshall Bullitt Mathematics and Astronomy Collection, Department of Rare Books and Special Collections, The University of Louisville; The E. P. Dutton and Company Records, Syracuse University Library, Special Collections Department; The Archives of American Mathematics, Center for American History, The

University of Texas at Austin (letters concerning Bell's MAA presidency); The Harry Ransom Humanities Research Center, The University of Texas at Austin, Austin, Texas (letters to Glenn Arthur Hughes).

For permission to quote from other individual letters and unpublished material, I am indebted to most of the above named archives as well as to the Special Collection, University Library, University of California, Santa Cruz; The Manuscripts and University Archives, The University of Washington; Special Collections, The Brown University Library; and The Office of the Home Secretary of the National Academy of Sciences.

I have also been fortunate in having the indispensable assistance of the public libraries of New York City, Palo Alto, Payette, Peterhead (Scotland), San Francisco (in particular, its very complete McComas Collection of Fantasy and Science Fiction), San Jose, Seattle, Upper Norwood (Surrey, England), and Yreka as well as the Library of the University of Aberdeen.

Bell's letters and unpublished writings are quoted by permission of Dr. Taine T. Bell; Muriel Rukeyser's letters, by permission of William L. Rukeyser; and the entries from Grace Hubble's journals, with the permission of The Huntington Library, San Marino, California. The line "imagination's other place" is quoted by permission of James Kirkup from his book, *A Correct Compassion*, Oxford University Press.

Quotations in captions are from Bell and the title for Part One, "the eloquence of facts," from Henry Adams.

I am greatly indebted to Andrew Underwood's *Bedford Modern School of the Black and Red*, Charles M. Gates's *The First Century at the University of Washington*, and Judith Goodstein's *Millikan's School* as well as to the many books on science fiction I have cited in the text.

The manuscript has been read at various stages by D. J. Albers, G. L. Alexanderson, Tom M. Apostol, Katherine Brody, R. P. Dilworth, Lincoln K. Durst, Neil D. Reid, Douglas Robillard, and R. M. Robinson. I very much appreciate their comments and suggestions. It goes without saying (but one must of course say it) that any errors remaining are most certainly my own.

Lastly, although I have already described the circumstances of the generous funding of my research, I wish to express here my thanks to James Vaughn Jr. and the Vaughn Foundation.

Contents

ONE

the eloquence of facts

A Cat That Can't Be Caught

Half a century after the publication of Eric Temple Bell's *Men of Mathematics,* I find myself driving down to Watsonville, California, a little town on the coastal plain of Monterey Bay, to meet the only son of its author. My companions on the drive are D. J. Albers of Menlo College and G. L. Alexanderson of Santa Clara University. Seven years earlier, at Albers's suggestion, they began work on a biography of Bell. Since 1981, off and on, they have been collecting material. They have talked to Bell's son and daughter-in-law, who are physicians in Watsonville, and have made trips to Southern California, where Bell—a distinguished number theorist—taught for many years at the California Institute of Technology. They have interviewed mathematicians and others who knew Bell and gathered what published material there is about him. But over the years they have become increasingly involved in professional activities. It is clear that they are not likely to have time very soon to complete their project. I have suggested that they write up the material they have as a dedicatory profile in the sequel to their book, *Mathematical People,* a collection of interviews and profiles of contemporary mathematicians.

On the drive down to Watsonville, they are quite frankly trying to talk me into writing the profile I have proposed.

Although I am not a mathematician, I know personally of the inspirational effect of Bell's classic. When *Men of Mathematics* appeared in 1937—lively, opinionated, its author writing about mathematics and mathematicians in a way that no one had ever written about them before—my younger sister, Julia Bowman, was a mathematics major at a small college in Southern California, essentially a teachers' col-

3

lege. Many years later, as Julia Bowman Robinson, she became the first woman mathematician to be elected to the National Academy of Sciences and the first woman to serve as president of the American Mathematical Society. As a college sophomore she read *Men of Mathematics* the year it came out. I don't know how she happened to hear about it so quickly, but it was through Bell's book that she first realized there was something she could do with mathematics besides teach it. She could *become a mathematician.* Just before her death in 1985, going back over her career, she said to me that she could not overemphasize the importance of such books about mathematics in the life of a student like herself who had been completely out of contact with creative mathematicians.

Under such circumstances I am not adverse to giving Bell his due as the father of mathematical biography, especially since most of the research seems to have been done by Albers and Alexanderson.

"O.K.," I tell them, "I'll write something."

In the course of our trip to Watsonville, the two of them summarize for me the principal facts of Bell's life, so far as they know them. Born in Scotland. Educated in England. Came to the United States at nineteen. A.B. at Stanford, A.M. at the University of Washington, Ph.D. at Columbia. On the faculty of the University of Washington from 1912 to 1926; on the faculty at Caltech from 1926 until his retirement in 1953. A highly respected number theorist, who published more than two hundred research papers. "Two hundred?" "That's right." Author also of a number of popular books on science as well as sixteen science fiction novels and two volumes of privately published poetry.

"A very colorful character," they emphasize. "There are *lots* of good stories about him."

But a fairly standard academic life, so it seems to me, except for the juxtaposition of mathematics, popularization, science fiction, and poetry, which *is* an unusual combination.

Bell's only son, Dr. Taine Temple Bell, a retired internist, and his wife, Janet, a practicing pediatrician, live outside Watsonville, a short distance from the hospital where Bell spent his last months. Their house, a low rambling structure that has been enlarged at various times to meet the needs of a family of three now grown children, sits on a gentle knoll. The area around the house is agricultural. There are hedges of raspberries across the road, and beyond them an apple orchard, which must be lovely when it is in bloom. A wooden gate opens onto a small courtyard and introduces the oriental feeling that pervades house and garden.

We are greeted at the door by Dr. Janet Bell, a friendly but no-nonsense woman whom one would like to have taking responsibility for the health of one's children. She leads us through the house to the living room. There Albers and Alexanderson are greeted warmly, and I am introduced to Dr. Taine Bell, who has had bypass surgery since they last talked to him. I expect Bell's son to look at least a little like his father—the familiar aquiline profile, the thick eyebrows, the striking eyes, but I do not see any physical resemblance. Taine Bell is taller, probably larger. He used to be a cross-country runner in his school days. He seems to be a kind man, genuinely interested in meeting me as a person, not simply as a potential biographer of his father. I think he would be a good doctor.

I am surprised at the degree to which E. T. Bell is still present in his son's home. It would be incorrect to call the house a shrine to his memory, but there is evidence of him almost everywhere one looks. On the walls are pictures painted by him and his wife, who was an art teacher before her marriage. His pictures are large and vivid, usually of desert locations. Hers are smaller, much less colorful. On chests and shelves throughout the house are pieces of Chinese porcelain and cloisonné that he loved and collected. His books—particularly the science fiction novels with their flashy covers—are easily picked out on the bookshelves. He had a fondness for putting a color-word in a title, and I note *Green Fire, The Purple Sapphire, The Gold Tooth*. I have to search more carefully for *Men of Mathematics, The Development of Mathematics, Queen of the Sciences*. The Bells point out to me that some of his books still contain the paper napkins, eucalyptus leaves, and matches with which he customarily marked his place. At Alexanderson's suggestion, they take down a volume of great interest to mathematicians, especially to those who are number theorists, as Bell was. A gift from friends and colleagues at the time of his retirement, it is the famous 1670 edition of the *Arithmetica* of Diophantus of Alexandria—"the humble source of the theory of numbers," as it has been called. This edition of the Bachet translation was republished by the son of Pierre Fermat after his father's death in 1665. It contains in its margins the notes that the French lawyer/mathematician—"the Prince of Amateurs"—made in his personal copy of Diophantus, including that most famous one relating to what is known as Fermat's Last Theorem: "[Of this] I have discovered a truly marvelous demonstration which this margin is too narrow to contain."

I am impressed by the direct simplicity with which Taine Bell speaks of his father. He is frank about the fact that it was not easy being the son of E. T. Bell. He tells us that even his own conception was an unwelcome surprise to his parents, whom both he and his wife refer to as "Romps" and "Toby." They were an exceptionally close couple—"for her everything was for him," he explains quite simply. He thinks they feared that a child would come between them. He speaks, again simply and directly, of how his father tried to push and pull him into academia. There were no summer vacations for him as a boy. Instead there were lessons in algebra and geometry from his father. "Brains were always the commodity," he says. Up to the end of his life his father made no secret of the fact that he was disappointed in his son. An M.D. simply did not rate with a Ph.D.

Taine Bell is very interested in having someone write an account of his father's life. Although Bell died in 1960 there has not been even the usual summary of the life and work of a departed member published in the *Biographical Memoirs of the National Academy of Sciences*. Two men in succession began to write the NAS memoir, but both died before having got even fairly started. Taine Bell is clearly disappointed by the fact that Albers and Alexanderson are not going to go through with their plan for a full-length biography. He is happy, however, to cooperate with me, as he has with them, by sharing what he knows. But he also explains, apologetically, that he does not know very much about his father's early life. He describes the elusiveness of his father, his unwillingness to be pinned down to anything about himself, even something as small as how he lost the last joint of his right thumb.

Janet Bell supports what her husband says. She first knew her future father-in-law when she and Taine were medical students at Stanford. It was a time when relations between father and son were particularly bad, and she remembers occasions on which Bell castigated Taine in front of others for his choice of a profession. Yet "Romps" could be very charming, she says. He was always attractive to women. "Intelligent, good-looking, interesting women," Taine Bell adds. He recalls the excitement among the librarians at Caltech when on one of his own visits back to Pasadena they heard that "Dr. Bell" was coming to the library and thought it was to be his father.

During the last year of Bell's life, the year that he spent in the Watsonville hospital, Janet Bell became better acquainted with him. It was

she who at his dictation took down a vague genealogy and also identified the memorabilia he had brought with him from Pasadena. It is a
magpie kind of collection. There is his mother's "lap desk" with a little
embroidered folder:

> England's letters find no grace
> Unless they bear Victoria's face.
> To guard her head from dust or damps
> This case provides for postage stamps.

The desk now contains the various school caps Bell saved from his days
at the Bedford Modern School.

The mention of Bell's school days in England brings up a question
I have been waiting to ask. Why had he, at the age of nineteen, emigrated to America and enrolled at Stanford University—at that time a
school that had been in existence little more than a decade?

The Bells do not know.

"We think that his father may have come to San Jose," Janet Bell
says, "but we don't know whether Romps came to California to find
his father, or his father followed him to California."

I am surprised at the vagueness with which she conveys this information since San Jose, like Stanford, is only a relatively short distance
from Watsonville.

She admits to vagueness. It happens, she explains, that at virtually the same time Albers and Alexanderson first talked to her and her
husband about writing a biography of Bell, their son, Lyle, decided to
do some research into the family's English roots during a business trip
to England. In Bedford he was able to locate the grave of Helen Jane
Lindsay Lyall Bell, his great-grandmother—the mother of E. T. Bell.
There was no longer a stone on her grave, but the sexton informed
him that the cemetery had at one time made copies of all "kerbstone"
inscriptions and hers was probably among them. He was too busy at
the moment to look it up, but he promised to telephone Lyle later at his
hotel. That evening he called, as promised, and read the inscription:

> Helen Bell
> Widow of James Bell
> Of San Jose, California
> August 29, 1924

Lyle Bell could not believe what he had heard, and he asked the
sexton to repeat. But there was no mistake. "Widow of James Bell of

San Jose, California." It was the first time that he, or any other member of the Bell family, had ever heard anything to suggest that the father of E. T. Bell had also been in the United States—and, of all places, in San Jose. The information was even more bewildering to the family because, as far as they knew, Bell had never had any contact with his father after he himself left England at nineteen.

After Janet finishes the story of Lyle's discovery, Taine volunteers that his mother once told him that his father's father had been "the owner of a fleet of sailing vessels," that his father as a small boy had been forced to climb the rigging and do other such things around the ships, all of which he hated, and that ultimately he had been "shipped out" by his father on a voyage that took him around the Cape of Good Hope to India and China. Taine adds that it has always been his impression that his father emigrated to America to escape from his father and from having to go into his father's shipping business.

"Your mother told you about the voyage to India and China?"

"Yes."

"Did your father ever make any reference to such a voyage?"

"No."

In fact, his father never spoke at all about his youth and hardly ever about his family. Taine Bell knows that he was the youngest of three children, the others being Enid Lyall Bell and James Redward Bell. Enid's middle name was clearly the mother's maiden name. Taine does not know the derivation of "Redward," but Janet suggests that it may have been the maiden name of the paternal grandmother.

"And what about Temple?"

Taine Bell, whose own middle name is Temple, has no idea.

"It was my father's name."

He has never had any contact with his English relatives. Up to the end of Bell's life, he was firm in his instructions that if he were ill or dying, or even after he was dead, under no circumstances was his son to contact the English family.

"Did you ever ask your father why?"

"No. I couldn't do that. I was not supposed to. It was drilled into me by my mother. I don't know if she actually said that to me, but I got the idea. I was never to ask my father anything about himself."

Janet now brings out a box of photographs that Bell brought with him to Watsonville and that Albers and Alexanderson have seen before. There are a number of pictures of Bell and his wife, as well as pictures of Bell's mother and his sister. There is also a very old photograph of a number of people posed in front of a large and imposing brick house. On the back the house is identified as "Penybryn, our London home. My father, born 1837, seated at the right."

This is the only picture of the father.

By this time I am beginning to think that the life of Eric Temple Bell may not have been so straightforward as Albers and Alexanderson have led me to believe. But it is late. From the Bells' living room we can see the sun going down behind the apple orchard. We all shake hands, and I tell Taine again, to be sure that he understands, that I will write merely a kind of dedicatory profile of his father for the sequel to Albers and Alexanderson's *Mathematical People*. I explain that the book will contain a number of interviews with outstanding contemporary mathematicians.

"You would never have been able to interview my father," he says with a wry smile. "He would never have let himself be pinned down."

"A cat that can't be caught?" I suggest, referring to Marianne Moore's description of T. S. Eliot.

"That's right."

NAME (Print) _____Bell_____ _____Eric_____ _____Temple_____
                    Last Name   First Name   Middle Name

DATE AND PLACE OF BIRTH _7 Feb. 1883, Aberdeen, Scotland._

IF OF FOREIGN BIRTH, DATE, SHIP AND PORT OF ENTRY INTO THE UNITED STATES _June 1902 j Summer_
Washington (S.S. Tunisian to Montreal, Canada, June 1902)

DATE AND PLACE OF NATURALIZATION (HOW OBTAINED) _3 Nov. 1924, Superior Ct. King Co., Seattle, Wash_

FATHER'S NAME AND BIRTHPLACE _James Bell, London, England_

MOTHER'S MAIDEN NAME AND BIRTHPLACE _Helen Jane Lindsay Lyall, Aberdeen, Scotland_

WIFE'S MAIDEN NAME (OR HUSBAND'S NAME) _Jessie Lillian Smit (widow Brown when I married)_

BIRTHPLACE AND PRESENT EMPLOYMENT _Deceased_

EDUCATION—DATES OF ATTENDANCE; NAMES OF SCHOOLS OR COLLEGES AND LOCATION:
Print

Common School: Name _Orkney House, Bedford, Eng_   No. of years completed _6 (all)_

High School: Name _Bedford (Eng.) Modern School_   No. of years completed _4 (all)_

Business School: Name   No. of years completed

College: Name { _University of London, Eng._   No. of years completed _3 (all) [A.B]_
                       Stamford Univ.   _[A.M]_
                Name { _Univ. of Wash._   _[Ph.D]_
                       Columbia University

10

An Unanticipated Question

When I begin to write an account of someone's life, I always think—although I should have learned better by now—that it is going to be easy. I have a sense of the life, the spirit of it, and I have at the moment confidence that I can catch that. Details will come later. At the beginning I want to sit down and hammer out a draft. In the case of E. T. Bell, in spite of certain questions I have, I think that the writing should be relatively easy. What I am to write is to be brief, and much of the research has already been done by Albers and Alexanderson.

A few days after the visit to the Bells I rough out some thirty typewritten pages. The result surprises me. I haven't expected what I write to be final, but neither have I expected it to wobble on its factual foundations. Of course, there is that vague story of the youthful voyage to the Orient. It seems curious that Bell never mentioned such an experience in writing about himself. And then there is Lyle Bell's discovery that his great-grandfather—E. T. Bell's father—also came to America. It seems curious that Bell never mentioned *that* fact to his son. But if Bell's father came after Bell himself did, given their reputed alienation, perhaps his passing over the fact is not so surprising.

Taine and Janet have given me copies of various writeups on Bell before and after his death as well as some autobiographical forms that he filled out at various times in his life. There is also the brief genealogy that Janet took down at his dictation during his last year. I decide that the first thing I should do is to correlate the information in these miscellaneous documents.

E. T. Bell was born. Where? In Scotland. But on some occasions, as in *Twentieth Century Authors*, he said in Aberdeen, on others—mainly on official forms—he said in Peterhead. Biographical writeups by others are equally contradictory about his birthplace. A small matter, I decide. After all, Peterhead is a little town in northern Scotland, about thirty miles up the eastern coast from Aberdeen, as I discover in the *Encyclopædia Britannica*—virtually a fishing village at the time of Bell's birth and for many years thereafter, although now the center for North Sea oil. Who in the United States during Bell's day would have heard of Peterhead? I assume that it was his Scottish mother's native town and that she returned there to have her child. Still, I have to know. Was his birthplace Aberdeen, or was it Peterhead?

Then there is the question of Bell's parents: his father, who came "of a prominent mercantile family in the City of London," and his mother, whose people "were classical scholars for several generations back." These are Bell's own quoted words in the entry for him in *Twentieth Century Authors*. They were later picked up by the mathematical historian Kenneth May for his writeup on Bell in the *Dictionary of Scientific Biography,* and repeated by other writers. But Taine has told me (as his mother, he says, told him) that his father's father was the owner of a fleet of sailing vessels. Why had Bell not stated specifically in *Twentieth Century Authors* the nature of his father's business?

This question brings up another. What about his father's packing him off to sea at an early age? According to Taine, it was to escape the sea and his father's shipping business that Bell emigrated at nineteen to the United States. Why would a man like Bell, writing about himself, pass up such an adventure as a voyage to the Orient? Instead, describing his background and early life for *Twentieth Century Authors,* he devoted more than two-thirds of his first 250 words to having lived "at one time" on a hill overlooking the old Crystal Palace Grounds where there were a few restorations of dinosaurs grouped around a small lake—"the beginning of a lifelong interest in those magnificently amoral brutes." Not a word about having gone to sea as a boy!

This omission seems so odd that I decide the first thing I must do is to try to determine from the information I have on Bell's life *when* that voyage to India and China could possibly have taken place. Of the several mimeographed forms that Bell filled out at various times for the administration at Caltech, the one that has the most detailed information about his early years is dated 1943, when he was sixty. The information about his schooling on this form is more detailed than that in his statement in *Twentieth Century Authors,* where he wrote merely, "My education was by tutors till I entered the Bedford Modern School."

Here he noted very specifically that he spent six years at Orkney House, a private preparatory school in Bedford, England, and four years at the Bedford Modern School, a private secondary school. He emphasized in both cases that these years were "(all)" that were offered at the schools. He also listed one year at the University of London but did not add "(all)" in that case.

This information provides me with a framework in which I can perhaps place the long ocean voyage that he described to his wife and that she described to their son. The educational record, as detailed on the form, accounts for eleven of his early years: one at the University of London, four at the Bedford Modern School, six at Orkney House. If he entered the preparatory school at seven, which would have been customary, then eighteen of the nineteen years that preceded his emigration to the United States are accounted for. He could have "shipped out" between leaving the Modern School and entering the University of London, or perhaps even earlier. A boy's going to sea at twelve or thirteen at that time would not have been unusual, so an English friend assures me: "The family was *in trade,* you see. So he would not have gone to sea to learn to sail, but to learn the *trade.*" Reasonable. But, still, why had he himself never mentioned the voyage to his son?

While I am mulling over this mystery, I am also going through other papers that Taine and Janet have given me. One of these is a reprint of Bell's memoir of Edward Mann Langley, the mathematics master at the Bedford Modern School whom he credited (again in *Twentieth Century Authors*) with "converting" him to mathematics. In the first paragraph of the memoir I come upon a sentence that stops me abruptly:

"Every detail of his vigorous, magnetic personality is as vivid today as it was on the afternoon I first saw him, thirty-five years ago," Bell wrote, ". . . as a new boy in the Upper Fifth."

I read the paragraph again. *As a new boy in the Upper Fifth.* There is no mistaking. Bell is saying that he entered the Bedford Modern School in the middle of the fifth form. I know enough about the English education system to know that the sixth form is the highest and the last. The fifth is next to last. If Bell entered as a new boy in the upper fifth, how could he have remained at the School for "4 years (all)"—as he stated in 1943 on the form that he filled out at Caltech?

He wrote of Mr. Langley with great respect:

"It was impossible for him to deceive himself as to the validity of a proof. An assertion was either proved or it was not, and Mr.

Langley never descended to the shabby trick of passing off as proof a tissue of hidden assumptions which might easily but falsely convince immature minds and, possibly, satisfy a professional examiner. If a chain of reasoning rested on demonstrable assumptions beyond the complete comprehension of the class, he said so, emphatically. Later on, a closer approach to the central subtlety would be made, and sometimes the difficulty would be disposed of for good. But no student of Mr. Langley's with normal intelligence ever left him with the delusion that something had been proved when it had not."

Reading this paragraph, I know that while Bell might obfuscate the details of his education for the administration at Caltech, he would never permit himself an untruth when writing about Mr. Langley. If he entered the Bedford Modern School in the middle of the next to last year, he would say so. But why had he *not* said so when summarizing his education for the administration at Caltech? And why had he emphasized that the years he spent at Orkney House and at the Modern School were "(all)" that were offered at those schools? In fact, I find it extremely odd that with an A.B. from Stanford, an A.M. from the University of Washington, and a Ph.D. from Columbia University, he would have even considered it necessary to list his elementary and secondary education—and in such detail.

I recall that at Taine and Janet's I saw a number of prize books that Bell received during his English school days. At the time I hadn't looked too carefully at them, but I did note that some of them bore dates. Now it occurs to me that to establish an accurate chronology of Bell's early education it may be helpful to go over those inscriptions.

When I arrive in Watsonville, two weeks after my first visit, I find that Janet has prepared for me by spreading out on the dining room table all the books that Bell brought with him to Watsonville. In the course of this activity she has begun to examine the inscriptions herself, and she has come across something that she is eager to show me.

"Look at this."

She hands me a leather-bound volume that looks like a schoolbook. There is a page marked. I glance at the spine of the book, *Commentarii de Bello Gallico,* and then open it at the marker, which is at the title page. There the title is in English—*Caesar's Commentaries on the Gallic War.* In the center, above the title, "James Bell, Central Hill, Norwood" is written neatly, but the rest of the page is covered with the scrawl of a boy's writing his own name over and over. There is also, at the bottom of the page, the name "Temple"—but in a different ink and printed rather than in script. Perhaps, I think, it is the name of a school chum. A few pages over, Janet points out, there is a signed notation in pencil by E. T. Bell: "This was one of my father's schoolbooks (at age 13)." But these things are not what has excited Janet. Turning back to the title page, she points to three inscriptions at the very bottom. Also in pencil, they are so light that one could easily pass over them:

Jan. 4, 1915	Jan. 4, 1921	Jan. 4, 1926
8:00 p.m.	8:15 p.m.	7:58 p.m.
In mem. ETB	In mem. ETB	In mem. ETB
	25 yrs.	30 yrs.

I do a little quick subtraction.

The death memorialized would have occurred on January 4, 1896.

"How old would Romps have been at that time?" Janet asks.

"Twelve," I reply promptly, prepared by my reading about Bell's life in the past fortnight. "A month short of thirteen."

"What do you suppose they refer to?"

"They could refer to his father's death."

"Possibly in San Jose?"

"Quite possibly."

Taine listens but does not say anything, only shakes his head from time to time.

Janet and I then begin to go through the other books and materials that Bell saved from his days in Bedford. Among these is a citation of Eric, his brother, and a friend for reckless riding of their bicycles: "for that you, on the Tenth of May in the year of our Lord One Thousand Eight Hundred and Ninety Six at Clapham in the County aforesaid [Bedford] did unlawfully and maliciously break a certain fence there . . . doing injury to the amount of six pence." There is just one book from Orkney House, a "First Prize" titled *Stories of Old Renown, Tales of Knights and Heroes.* It bears the date April 1897. There are half a dozen or so prize books from the Bedford Modern School. The earliest of these is *Poetical Works of Alexander Pope,* a prize for English dated Midsummer 1898. Of the other prize books none bears a date later than 1900, two years before Bell emigrated to America.

Among other books Bell kept, several belonged to his father, James Bell Jr.—which is how he wrote his name in most of them.

(At this point I realize that I am dealing with just too many people who are named Bell. From now on, I shall refer to the present members of the Bell family only by their first names—Taine, Janet, and their three

THE BELL FAMILY

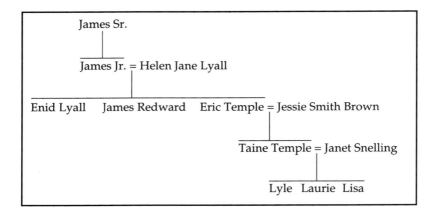

James Sr.

James Jr. = Helen Jane Lyall

Enid Lyall   James Redward   Eric Temple = Jessie Smith Brown

Taine Temple = Janet Snelling

Lyle Laurie Lisa

children—Lyle, Laurie, and Lisa. When I write *Bell* I shall be referring to E. T. Bell. His brother, James, will be James Redward, his father James Bell Jr., his mother Helen Bell (although she most commonly wrote out her full name, Helen Jane Lindsay Lyall Bell), and his grandfather—if he comes into the story—James Bell Sr.)

Among the books belonging to James Bell Jr. I note a copy of *Hypatia*. This is Charles Kingsley's story of "the Divine Pagan"—the daughter of Theon the Mathematician—a beautiful woman and a brilliant lecturer on mathematics, who was murdered in Alexandria in the early fifth century by a mob of fanatical Christians. Below James Bell Jr.'s name on the title page Bell inscribed his own name and, again, the date—January 4, 1896—as if the book became his book on that day.

This last inscription seems to settle any question that has been in our minds as to whose death was memorialized by the three "In Mem." notes penciled on the title page of the Caesar.

There is something else that I have noted during the afternoon. Among all the memorabilia from Bedford that Bell kept, there is not one book or one piece of paper that bears a date *earlier* than January 4, 1896. The citation for reckless riding of a bicycle: May 1896. The Orkney House prize book: April 1897. The earliest prize book from the Bedford Modern School: Midsummer 1898.

I am already making plans to write the next day to the Santa Clara County Recorder's Office and find out if it has a death certificate for a James Bell dying in San Jose on January 4, 1896. If the answer turns out to be in the affirmative, it will bring up a very interesting possibility.

At the end of the afternoon I find myself asking Taine a question I could not have imagined asking at the beginning:

Is it possible that his father, E. T. Bell, spent a portion of his childhood in San Jose, California, and in 1943, when he was sixty years old, a distinguished mathematician and a famous writer, felt it necessary for some reason to hide the fact?

Taine, who seems somewhat dazed by the various discoveries of the afternoon, does not answer immediately.

"He was very fond of San Jose," he says after a rather long pause. "He always stopped there on his way north. We thought it was because he had gone to Stanford."

LYLE BELL, the first grandchild, born 1949

What Lyle Had Learned

That afternoon, when I leave the home of Taine and Janet Bell, one thing is clear: the early life and family background of E. T. Bell had not been as he represented it throughout his life to his son and his son's family. The discovery by Bell's grandson, Lyle, of the inscription on the grave of his great-grandmother—Bell's mother—had opened a crack in the wall that Bell had constructed around the years of his youth. Why did he never mention his father's presence in San Jose? Why the careful, overly specific delineation of his English education on a mimeographed form? I do not know, but I do know that if I am going to write about Bell's life I have to try to find out what he walled off and why.

I must admit I have some misgivings about what I am going to be doing. Taine, who has been so honest and direct with me about his relationship with his father, is eager to have a long overdue biography written and is clearly disappointed that I have promised only a profile. In the years, almost thirty now, since his father's death he has exhibited a strong sense of filial responsibility in cooperating with publishers to keep Bell's works in print. He has also made extensive, although so far unsuccessful, efforts to get into print a long poem that Bell always considered his masterpiece. I can't help wondering how he and the rest of the family may react if my discoveries turn out to reflect unfavorably on the famous progenitor of their family.

I also have qualms, I must admit, about exposing something in Bell's life that he himself, clearly, did not want others to know.

Yet the chase is absorbing, and I do not want to turn back.

The day after my trip to Watsonville I dispatch a letter to the Office of the County Recorder of Santa Clara County inquiring whether there is a death certificate for a James Bell Jr. dying in the city of San Jose on January 4, 1896. I also send off a letter to the Modern School in Bedford, England, requesting a copy of the school record of Eric Temple Bell. I cannot expect a reply to either of these requests for four to six weeks, but I am still able to move ahead. Janet has promised to put pressure on her son, Lyle, to send me a report of his 1981 genealogical researches.

"I have the reputation in the family of being the one who gets things moving," she says.

She is quite right. Almost immediately I receive a bulky mailing from Lyle with a detailed summary of what he had learned about the Bell family in England.

Lyle is Bell's only grandson and the eldest grandchild. He attended Claremont Men's College on a five-year "co-terminal" program with Stanford University that combined economics and engineering. At the time of his researches in England, he was Director for Management Information Systems for Upright, Inc., a small company that manufactured portable aluminum scaffolding and scissor lifts and had a factory in Ireland. He is now Assistant Vice President for Data Processing at Caesar's Palace in Las Vegas.

The material that he sends me contains several new and interesting pieces of information.

One of the places he had been most eager to visit while in England was Penybryn, "our London home," as Bell described it on the back of the photograph that I have been shown by Taine and Janet. The photograph is of a tall brick house, the top of which is obscured by the branch of a great oak that frames the picture. There is a spacious terrace at the side of the house from which the photograph was taken. A number of people, including two children, are posed on the terrace and on the steps leading down to what is apparently a sloping lawn. Others look out from upstairs windows. On the back of the photograph Bell identified the youngish, bearded man at one end of a loveseat constructed of woodsy branches as his father, but a white-haired and white-bearded man, seated in the window above the terrace, a kind of presiding figure, dominates the scene. Bell did not identify him, but I am quite certain that he is James Bell Sr., the grandfather.

Lyle did not have a street address for Penybryn, only the name. He knew though, from what Bell had told Taine, that the house was located in Upper Norwood, a suburban district that lies south of London proper in the borough of Croydon and the county of Surrey. Before he left for England, he wrote to the librarian of the Central Library in Croydon to try to obtain a current address for Penybryn. The librarian replied that the name Penybryn still appeared on the most recent Ordnance Survey Map of the area, which was, however, twenty years old. The house was listed there as No. 15 Fox Hill and was, according to the then current Electoral Roll, still occupied.

"Unfortunately, Upper Norwood has had a great deal of redevelopment and without a personal visit, it is impossible to be sure that the present No. 15 is the old *Penybryn*."

Thus it was that, on a rainy May weekend in 1981, Bell's grandson caught a train at Victoria Station and rode out to the Crystal Palace Station. A brisk fifteen-minute walk took him to Fox Hill. There at the top of the hill, on a corner, almost hidden by trees, was Penybryn, the house of the photograph.

Penybryn—a Welsh name meaning "the chief of the hill"—is located at the intersection of Fox Hill and Tudor Road. It is a big, tall house, originally red brick and ornamented with diamond-shaped vertical chains outlined in black brick—a motif that still appears on most of the brick houses in Upper Norwood. By the time of Lyle's visit, however, the front and one side had been stuccoed over and the wide double door painted bright red (a defacement unfortunately common in the area). Nevertheless, Lyle found the house still imposing, although both it and the surrounding garden had been allowed to run down. The terrace on which Bell's father had been photographed was still there, also the long sloping lawn. The oak that had been large even at the time of the photograph now completely obscured the famous view of the valley below, but the valley could still be seen from the middle of the intersection of Fox Hill and Tudor Road.

No one answered Lyle's knock, but an elderly neighboring couple saw him wandering around the neighborhood in the rain and invited him in. They gave him a drink and told him what they knew about the house. It was currently occupied by art students. In fact, it had something of a place in the history of art, for at one time it had been painted by Camille Pissarro, who had lived in Upper Norwood during the German occupation of Paris in 1871. The painting could be seen in the National Gallery in London, they told Lyle—they thought that it was called "Red and Black House," but it is in fact titled "Fox Hill, Upper Norwood."

After his visit to Upper Norwood, Lyle went to Bedford, both that weekend and on a later occasion during the week when public offices were open. There, as I already know, he discovered the totally un-expected inscription that had been on his great-grandmother's grave. He also visited the offices of the Modern School and obtained a copy of the sheet upon which the enrollment of Eric Temple Bell had been recorded. Bell enrolled in January 1898 and left in May 1900. I note that these dates are consistent with his having entered the Modern School in the upper fifth form.

From the bicycle citation of 1896, Lyle had an address for the Bell family in Bedford. After seeing Penybryn, "which was *really* impres-sive," he was shocked to find that the residence in Bedford was a very modest two-family house. It was clear that the move from Upper Nor-wood must have been the result of a drastic change in the family's financial circumstances.

Below are some relevant excerpts from comments that Lyle sends me along with notes and letters pertaining to his 1981 research. I am printing them here because they show the state of the Bell family's knowledge about E. T. Bell and his parents and siblings at the time I begin my research:

1876   Records show that James Bell [Lyle is referring here to James Bell Jr., the father of E. T. Bell] began occupying the large home called Penybryn in Upper Norwood in 1876. In 1876, Helen [James Bell Jr.'s wife and the mother of E. T. Bell] would have been 24 years old. Perhaps they were married in 1876.

1878   Enid Bell [E. T. Bell's sister] may have been born in 1878.

1880   James Redward Bell [E. T. Bell's brother] was born November 28, 1880. His birthplace is unknown.

1883   Eric Temple Bell was born February 7, 1883, in Helen's home-town in Scotland. It is unclear why he was born in Scotland. Perhaps James [Jr.] was on a voyage and wanted Helen to be taken care of by her family.

1889   Records show that Penybryn went from James Bell's name to Helen's name in 1889 or 1890. It appears that James left the family in 1889 and went to San Jose.

1896   School records show that James Redward [Bell] started in the Bedford Modern School in May, 1896.

1897   [This is] the last year that Helen appeared as owner of the
        house in Penybryn. Perhaps [James Jr.'s] death precipitated
        the move.

1898   E. T. Bell started at the Bedford Modern School in 1898.

It is Lyle's theory "that James and Helen were separated or di-
vorced sometime before 1896. I think he moved to America and E. T.
Bell went to America to try and locate his father." For Lyle, as he writes
at the end of the summary he sends me, the most intriguing questions
that remain are those relating to what James Bell Jr. did for a living in
England and why he emigrated to California—also what happened to
the older brother, James Redward Bell, from whom Bell cut himself off
so completely that there is not a single reference to him in all the Bell
memorabilia in Watsonville.

"I remember my mother saying that she had heard that Romps
did not respect his brother because he was involved in the slave-trade
business on the coast of Africa. James Redward [Bell] would have been
too young to be in the slave-trade business. I wonder, though, whether
the family was involved in slave trade earlier in the 19th Century. This
would be consistent with owning sailing ships. Perhaps James [Jr.] was
driven from England because of an association with the slave trade.
Perhaps he left something in California which would be a start for Eric
in some other business."

In spite of Lyle's interest in James Redward Bell, I am not immediately concerned with trying to track down a mysterious older
brother who is thought to have gone to someplace in Africa. I am
more concerned with why Eric Temple Bell, sixty years old, distinguished mathematician and famous writer, felt that he had to write
out in unnecessary detail an elementary and secondary school record
on a mimeographed form that he was completing for the administration at Caltech. It seems clear from what Lyle has written, based on
information he found in Croydon directories, that Bell's mother and
her three children remained in England after the father went to America, that the father continued to support them, but that upon his death
the support ceased or was so greatly diminished that they had to leave
the big house in Upper Norwood for a much more modest residence
in Bedford.

After I receive Lyle's mailing, I decide that I won't just wait for a
death certificate from the Santa Clara County Recorder's Office but will
try to find out if there was an obituary in one of the San Jose papers.

I telephone the California Room of the San Jose Public Library and ask the librarian, Barbara Dickinson, if she will check for a report of the death of a James Bell Jr. at the beginning of 1896. The next day I receive a letter with a copy of an item from the *San Jose Daily Mercury* of January 8, 1896:

> Bell—In San Jose, January 4, 1896, James Bell, a native of London, England, age 58 years and 11 months.
>
> Friends and acquaintances are respectfully invited to attend the funeral TODAY (Wednesday January 8th) at 10:00 A.M. from his late residence on Lucretia Avenue. Interment Surrey, England.

I know from Bell's note on the back of the photograph of Penybryn that his father was born in 1837. Penybryn, I know, is in Surrey. The identification can scarcely be more complete.

I immediately write to the Office of the Recorder of Santa Clara County and request "a copy of the will of JAMES BELL, who died in San Jose on January 4, 1896." Again I know I cannot expect to receive an answer for four to six weeks. In the interim I call the California Room, where Barbara Dickinson and George Kobayashi have been very obliging. Can one of them find a listing for James Bell* in the San Jose City Directories in the year immediately preceding his death?

Again, promptly the next day, I receive a reply. They have checked the directories, not just for the year before Bell's death, but for every year between 1870 and 1896. The earliest entry for a James Bell is in the directory for 1886–1887:

Bell, James, orchard 15 1/2 acres, Jackson, p o San Jose.

I am astonished. Is it possible that James Bell Jr., the owner of a fleet of sailing vessels came to California *to raise fruit*?

The entry on James Bell is followed by a series of letters: "re-pnnr-i-cwer-p-tur." The letters are a code, I am informed, the key for which was furnished to directory subscribers and was changed each year.

The library does not have a copy of the key for 1886–1887; however, from a key for a later year I find that *re* refers to real property and *pnnr* to its valuation; *i* to improvements and *cwer* to their valuation; *p*

---

* Since James Bell Jr. dropped the Jr. from his name after he came to California, I will follow suit in this and subsequent chapters where there can be no confusion with his father, James Bell Sr., back in England.

to personal property, and so on. If James Bell had had a mortgage on his property, it would have been coded with an *m*.

A copy of a second item from the the *San Jose Daily Mercury*, dated Tuesday, January 7, 1896, is also enclosed:

ESTATE OF JAMES BELL

R. E. Pierce filed a petition yesterday for letters of administration in the estate of James Bell, who died in this city on last Saturday.

The estate consists of sixteen and one-half acres of orchard land, about a quarter of a mile from San Jose,* valued at $8000 and brings in an annual income of $1000, personal property valued at $150, and $850 cash in bank.

"Now what does that mean?" I ask my husband, who is a lawyer.

"It means that James Bell died intestate, so you won't get a will," he says. "But obviously there was a probate, so you will find out something, although not so much as with a will."

Subsequent directory entries, which are also included in the mailing from the California Room, identify James Bell as "orchardist" and locate him more specifically "2 1/2 m SE of San Jose" and at "Lucretia av nr Story road." But this is not all I learn from the California Room's mailing. Almost as surprising to me as the discovery that Bell's father came to America to raise fruit is the information in the last paragraph of the letter from Barbara Dickinson.

"I added a page from the 1896–97 directory," she wrote, "because I thought it might be of interest to you."

There, underlined among the half dozen Bells listed in the year following the death of James Bell, is the entry:

Bell, Jas Mrs, moved to England.

If Mrs. James Bell, the mother of E. T. Bell, was the head of the household at Penybryn from 1889 to 1897—as Lyle wrote to me that she was—who then was Mrs. James Bell of San Jose, California?

---

* James Bell's acreage and its distance from town are different in every statement made about them.

HELEN JANE LINDSAY LYALL BELL, the mother of E. T. Bell

Who Was Mrs. James Bell?

It is beginning to appear that the parents of E. T. Bell will be as elusive as their son.

Although I now know that Bell's father was an orchardist in San Jose, California, I still do not know the business he left in Britain, or why he left it. And I do not know if the "Mrs. James Bell" who was with him in San Jose at the time of his death was the same Mrs. Bell who was listed in the Croydon directories—at the same time—as the head of the household at Penybryn.

While I wait for the material I have requested from the Recorder's Office, I decide to turn my attention to official records that are accessible, beginning with the English and American censuses. The United States Government takes a census every ten years—in the last year of the decade—so the census of 1890 can be expected to provide the names of the occupants of the household on Lucretia Avenue in San Jose in that year. The identity of "Mrs. James Bell" will then be settled by the first name, which will be given there. Her Majesty's Government—it was still Victoria's government during the first eighteen years of Bell's life—also takes a census every ten years, but in the first year of the decade. Its census of 1891 can be expected to produce the names of the occupants of the household at Penybryn in Upper Norwood for that year. I already know the whereabouts of Bell's father in 1890 and 1891—he was out on Lucretia Avenue two and a half miles from downtown San Jose, raising fruit. Through one or the other of the censuses I will learn the whereabouts of his wife and children.

I telephone the Federal Records Center in the San Francisco Bay Area, one of the twelve branches of the National Archives of the United

States, and ask if it has microfilms of the American censuses, specifically the census of 1890.

"The 1890 census? Sorry but there is no census for 1890."

"No census? How can that be? Isn't a census taken every ten years?"

"That's right. But the information collected for the 1890 census burned before it could be published."

Again I find myself repeating what I have just been told.

"Burned?"

"Yes. Almost entirely. There were a few districts—what district are you interested in?"

"San Jose."

I can hear pages being turned.

"No. I'm sorry. Nothing from San Jose."

Well, there is still the British census of 1891. I call the local Family History Library of the Latter Day Saints, more familiarly known as the Mormons, and inquire how to go about ordering the relevant portion of that census from the Family History Center in Salt Lake City. Again I am disappointed. Although the United States releases its census returns after seventy years, Great Britain holds its returns for a full century. Its 1891 census will not be released for several more years.

By this time I am literally obsessed with trying to unravel the mystery of the background and early life of E. T. Bell—and thoroughly frustrated. Everywhere I turn I find a question—and rarely an answer.

In due time my request for a copy of a death certificate for James Bell comes back from the Santa Clara County Recorder:

CERTIFICATE OF SEARCH
I hereby certify that I have searched the indexes of Santa Clara County for the year(s) 1873–1922 and I have not found a record [of death] for James Bell.
[Signed by a deputy for County Recorder]

How is it possible that there would not be a death certificate for a man whose death was reported in the local newspaper?

On a trip down to San Jose to examine the holdings of the Historical Museum, I look in the San Jose City Directory of 1896 for the name of R. E. Pierce, who filed the petition for Letters of Administration in the Estate of James Bell. I find him there: a physician with his offices in the palatial Hotel Vendome. (I can say "palatial," for although it no longer exists, there is a photograph in the office of Leslie Masunada, the historian at the museum, and it is truly palatial.) I assume that Pierce was James Bell's family doctor, probably a friend of the family as well, since he was asked by the widow to serve as executor of her husband's estate.

I learn more at the museum. Under glass on a counter I find a pictorial map of San Jose's "civic center" at the time of James Bell's death. I am surprised to see that the Records Office, where deaths were reported, is right next door to the Courthouse, where Dr. Pierce filed the petition for Letters of Administration. James Bell died on Saturday. Dr. R. E. Pierce filed the petition on Monday. Why did he—a medical doctor—not stop at the Records Office on his way to the Courthouse and report the death of James Bell?

It happens that at this time I am reading a novel about the murder of a mathematician and upon finishing it I say disgustedly to my husband, "I could write a better detective story about Bell!"

"Why don't you write a biography that is also a detective story?" he suggests.

"The problem is that I am not really interested in Bell the man," I explain. "I don't feel any empathy for him as an adult, at least not up to now, but I *am* interested in what happened to him as a boy and why he tried all his life to cover it up, whatever it was."

"Then just write about his youth."

Ah, *The Young E. T. Bell*!

Maybe the profile I am to write for the sequel to *Mathematical People* can focus on only the early years of Bell's life. I consult Albers and Alexanderson. It will be a good story, I assure them, besides being relevant to an understanding of the complex man that Bell the adult so obviously was. I propose ending the profile with Bell's coming to Stanford in 1902, although I am willing to go even a little further, say up to 1924, when he won the prestigious Bôcher Prize of the American Mathmatical Society.

Albers and Alexanderson, although polite, are a bit doubtful about such a truncated life. They, like Taine, are still hoping for a full-length biography. They do think, however, that I should try to go to Scotland and England to pursue my research into Bell's early life and family background, which they too find very interesting. Albers suggests that I may be able to get financial support for such a trip from the Vaughn Foundation. The trustee of that family foundation, James Vaughn Jr., has a well-known and longtime interest in Bell. In fact, after Bell's death he went to great lengths to have all Bell's unpublished manuscripts copied and made available to any library that wished to add them to its collection.

On February 4, 1989, I write a long letter to Mr. Vaughn, detailing my findings to date and explaining why a trip to Scotland and England seems necessary:

"The principal questions I would expect to answer on this trip are (1) the cause of James's death (his body having been returned to England for burial), (2) the nature and financial circumstances of his occupation before he left England, and (3) the reason for ETB's lifelong and highly emotional estrangement from his English family. This last will involve finding out what became of his mysterious older brother, who is said to have gone to Africa and become involved in the slave trade."

It is quite a lengthy letter, but I am trusting that Mr. Vaughn's interest in Bell is great enough to sustain it, and it is. The Vaughn Foundation agrees to support my proposed research trip to Scotland and England, and I begin to make plans to leave at the end of September.

In the meantime I decide to turn my attention to British records that are accessible in this country. The 1881 English census, which has been released, will provide me with information about the occupation of James Bell Jr. before he emigrated to the United States. Had he indeed been the owner of a fleet of sailing vessels, as Bell had said?

I am fortunate that my daughter, a geologist, lives in Salt Lake City, so I am able to avoid the time-consuming process of ordering the census microfilms from Salt Lake City through the local LDS library. Armed with the street address for Penybryn, she has no trouble locating the entry for James Bell in the 1881 British census. She makes a copy of the relevant page and mails it to me.

It is very exciting to see the name you are looking for in a copy of a document that is more than a hundred years old:

Fox Lane. Penybryn. James Bell.

The name is underlined twice on the census rolls to indicate the beginning of the listing of a household. I race across the line on which it appears. *CONDITION as to Marriage. Widr.* Widower? *AGE last Birthday.* 74. Seventy-four? The James Bell I am looking for was born in 1837 and so would have been forty-four in 1881. I realize that at Penybryn I have got, not Bell's father, but his grandfather, James Bell Sr.

There is, however, something significant in the entry. James Bell Sr. is listed as having been born at St. Mary-at-Hill in the City of London, and his occupation is given as "Merchant."

"My father came of a prominent mercantile family in the City of London . . . ," so Bell had written, truly, for *Twentieth Century Authors*.

But what kind of merchant?

Again with the assistance of my daughter, I am able to trace Penybryn back through the English censuses of 1871 and 1861. In both of these the occupation of James Bell Sr. is given specifically. He was a "Fish Factor."

Since I am not quite sure what "factor" means in relation to fish, I look up the word in the dictionary:

> Factor . . . [ME, fr. MF facteur, fr. L factor doer, fr factus] (15c)
> 1: one who acts or transacts business for another; as a: BRO-
> KER . . . b: one that lends money to producers or dealers (as
> on the security of accounts receivable).

The definition seems to fit the fish market well enough.

So now I know something about Bell's grandfather. But what about his father? I still do not know the occupation in England of the James Bell who died in San Jose.

Lyle had found a James Bell occupying Penybryn, but it has turned out to be the grandfather and not the father. Yet, according to Lyle, a couple of years after James Bell Jr. emigrated to California, his wife, Helen Bell, was listed in the local Croydon directories as the head of the household at Penybryn. How could that have come about? James Sr. could have died some time after 1881 and left Penybryn to James Jr., who then—for reasons yet unknown to me—could have left his family ensconced at Penybryn, financially secure, and set off for the new world. Some such scenario is the only one I can imagine to explain Helen's presence in Upper Norwood rather than in San Jose with her husband. But then how and why had "Mrs. James Bell" been in San Jose when he died? And where had young E. T. Bell been all this time? I am right back where I started. I can comfort myself only with the fact that in genealogical research, as in mathematics, a negative answer is an answer, too. I have at least learned that by 1861 Bell's father had left home. If he was not at Penybryn, *he was somewhere else.*

I now get another negative answer. My request for a copy of the will of James Bell comes back from the Recorder's Office with a yellow "postit" in the corner and a handwritten note from "Pattie," explaining that James Bell died without a will, as I already know, but that she has located the number of his probate file. Fine, I think, at least from the probate file I will be able to obtain the name of his widow, and that is what I am really after.

But there is a second page with a further note from Pattie. The number of James Bell's probate file is 3544, and all probate files from 3539 to 3590 *are missing.* "Sorry."

At this point the thought that a negative answer is an answer too is cold comfort. My husband, however, points out that even if the probate file is missing, the Clerk's Register should still exist.

"What is the Clerk's Register?"

"It's the clerk's record of all pleadings that take place in a probate. Fees paid, petitions filed, things like that. It won't contain the actual documents that would have been in the probate file itself, but it will contain a list of those documents as they were filed."

"How will I get to look at it?"

"Well, it just happens that I'm going to be taking a deposition in San Jose next week. If I have time I'll see what I can find for you."

The deposition in San Jose finishes early, and that evening over dinner my husband reports what he has learned during the course of the afternoon.

He went first to the Recorder's Office to make sure there was in fact no death certificate for James Bell. Since it was Friday, the clerks were not busy. When he explained what he was looking for, they gave him a smock and a pair of white gloves with which to handle the yellowed and crumbling pages of the actual book in which deaths at that time had been recorded—alphabetically, a page or so to each letter, and chronologically as they were reported. He found that it was true, as I had been informed by the Recorder. In spite of the announcement in the newspaper, the death of James Bell in San Jose on January 4, 1896, *was never recorded.*

He went then to the Probate Department, where the staff was equally cooperative. First he checked the probate records and saw for himself that an entire group of probate files, which would have included that of James Bell, was missing. He then asked to see the Clerk's Register. From it he obtained the information that I already had about Dr. Pierce. Of course what he really wanted to discover was the identity of "Mrs. James Bell." In this he was frustrated, for at every point where the widow entered the record (requesting, for example, "a family allowance") she was identified only as "the widow." He kept asking for additional documents. "There should be this." "There should be that." The four clerks, becoming increasingly excited by the search—I

know just how they felt—made suggestions and climbed up and down
ladders to get additional documents. But always the identification was
the same.

The widow.

Notice of the Sale of Real and Personal Property was filed in February, the sale took place in March, there was an accounting in April.

The widow.

"There may be an Order of Guardianship," my husband suggested
to the oldest clerk.

Up the ladder again, and down.

In the matter of the Estate and Guardianship: Enid Lyall Bell et al,
Minors.

"But the widow. Who was the widow?" I demand.

"I'll get to that," my husband says calmly. "So there was nothing
in the Petition for Appointment of Guardian, but there was still the
Decree of Settlement of Final Account. This old fellow, who was really
the most interested in what I was looking for, went running up the
ladder again and got that."

He places in front of me a practically unreadable copy of a legal
document that has been made from microfilm.

"You see, since James Bell died intestate, one half of what he had
went to his children, the other half to his widow."

He points to the bottom of the second and final page:

*"One half of said residue to Helen J. L. L. Bell, the surviving widow of
said deceased."*

There she was, the surviving widow—*in San Jose, California.*

But where were her three children?

PENYBRYN in Upper Norwood, Surrey, "our London home"

They Say, What They Say, Let Them Say

By now I am convinced that the only way I am going to determine whether E. T. Bell was in San Jose, California, before his thirteenth birthday, as I suspect he was, is by determining the whereabouts of his parents. They, however, continue to be elusive.

I have established that his mother was in San Jose at the time of or immediately after his father's death. But I have difficulty reconciling this information with Lyle's statement to me that, according to Croydon directories, "in 1888 or 1889 Penybryn passed from James to Helen." From the English census of 1881, which was not then available to Lyle, I now know, what Lyle did not, that the James Bell at Penybryn before 1888 was Bell's grandfather, James Bell Sr., seventy-four years old and a widower. How Helen—the wife of James Bell Jr. and the mother of E. T. Bell—happened to become the head of the household at Penybryn in 1888 or 1889 is a mystery I shall have to investigate when I get to England.

As to Bell's father, although I have him firmly settled in San Jose from 1886 to 1896, I have not yet been able to find him as an adult in England. He is not listed among the family at Penybryn in any one of the censuses of 1881, 1871, and 1861. In the earliest of these, when he would have been twenty-four, the family members are James Sr., his wife, a daughter, and another son, 18, also identified as a fish factor.

I decide that since I can't find James Bell Jr. in England, I will have to look for him in Scotland.

My Scottish research begins in an unexpected place. One afternoon, picking up Bell's classic *Men of Mathematics*, I am struck by the

curiously worded heading over the introductory section of quotations
on mathematics:

They Say, What They Say, Let Them Say.

The quotations are from a variety of authors that include—in
addition to familiar famous names in mathematics—such unexpected
names as Lincoln, Goethe, and the Mock Turtle. What catches my eye,
however, is the parenthetical identification of the source of the title
above them:

(Motto of Marischal College, Aberdeen.)

Marischal College is no longer listed among "Colleges and Universities of the World," but it occurs to me that an inquiry directed to
the University of Aberdeen may provide some information about Bell's
mother's family. According to what Bell wrote in *Twentieth Century Authors*, the family of his mother, Helen Jane Lindsay Lyall Bell, had been
"classical scholars for several generations back." It might be possible
to find in the faculty records of Marischal College some Lyndsays or
Lyalls, Lyles or Lindsays. (Names were not spelled consistently in the
nineteenth century, and Bell himself wrote both Lyndsay and Lindsay.)
A reply to my query comes promptly from Colin A. McLaren, Archivist
and Keeper of Manuscripts at the University of Aberdeen.

"I have been unable to identify with any certainty the Lindsays
and Lyles of Aberdeen to whom you refer. I note that a James Lyle, son
of James, farmer in Laurencekirk, attended the Arts Course at Marischal
College and University from 1832–36, subsequently becoming schoolmaster at Peterhead. Is he, perhaps, the founder of the dynasty?"

Peterhead—it seems quite possible that he is indeed. Shortly after receiving this information, I am in Salt Lake City and on several
mornings join the group of serious men and women armed with note-
books, briefcases, and file boxes who are already waiting in line when
the Family History Center of the Latter Day Saints opens. Because of
Mormon religious beliefs, which include the reunion of families after death, it is the goal of the Church to copy every record in existence that involves the names of human beings. The original copies
of the archives of the Center, which are stored in a vault in the granite walls of a canyon outside the city, have been accurately described
by Alex Shoumatoff as "a mountain of names." Signing the register as
instructed, I note that the archives are being used by many who like
myself are not members of the Church.

In the British Isles Section, which occupies an entire lower floor of the Family History Center, I am directed by a volunteer to the many indexes of Scottish births, deaths, and marriages. It is not difficult to trace James Lyle (more often Lyall) in these. A few years after he arrived in Peterhead he married a local girl, Grace Will, the daughter of a shipmaster. In due time they had three equally-spaced daughters: 1850, 1852, 1854. In my subsequent examination of the Peterhead censuses, however, I am not able to discover much about the middle daughter, Helen Jane Lindsay Lyall, who was to be E. T. Bell's mother—in fact, most of the time even her whereabouts are a mystery. There is no question but that she was born in Peterhead on October 6, 1852, the daughter of James Lyall, parish schoolmaster, and his wife, Grace Will Lyall. But in neither the 1861 nor the 1871 census of Peterhead is she or her younger sister listed as living in the Lyall household at 14 Prince Street. Her absence from the 1871 census is not too surprising, since in that year she would have been eighteen (the census was taken in April), but in 1861 she was eight and her younger sister six.

Having, however, established the fact that Bell's mother was born in Peterhead and had a family that continued to live there, I turn to the index of Scottish births to ascertain the birth dates of the children she had by James Bell. Working back in increments of two years from February 7, 1883, when E. T. Bell was born, I find James Redward Bell born in Peterhead on November 28, 1880, and Enid Lyall Bell born on August 6, 1878, also in Peterhead.

I am now in a position to search out the marriage of James Bell and Helen Lyall in the Index of Marriages. This index is arranged alphabetically by year, but with men's and women's names in separate listings, each name followed by the name of the parish in which the marriage took place. James Bell was not an uncommon name in Scotland in the nineteenth century, nor was Helen Lyall. The trick is to find a James Bell marrying a Helen Lyall and a Helen Lyall marrying a James Bell in the same parish and in the same year. But it does not take long.

James Bell, Cruden, October 18, 1877.
Helen Jane Lindsay Lyall, Cruden, October 18, 1877.

I know that I have the right couple thanks to Helen's insistence upon giving her entire name on every possible occasion or, at the least, all her initials. The next step is a letter and money order to the Register General, Search Unit, at the New Register House in Edinburgh, requesting the birth certificates of the three Bell children and the marriage certificate of their parents. I am counting on at least one of these documents providing information about the occupation of James Bell Jr. before he emigrated to America.

Impatient as always to *find out,* I decide that while I am at the Family History Center I will look for him in the 1881 census of Aberdeen on the theory that while Helen might have gone home to Peterhead to have her children, the family of the owner of a fleet of sailing vessels would probably have had its permanent residence in the larger town, which Bell usually gave as his birthplace. Going through the census of a city, or even a relatively small town, without the guidance of a street address or at least a parish, is not something that I would recommend. By the end of the day I have not found James Bell Jr. but, tired and glassy eyed from peering at reels of microfilm, I am not at all sure that I have not passed over him. Some weeks later, when I receive the requested marriage and birth certificates from Edinburgh, I find that I have been on a wild goose chase in Aberdeen. On all four of these documents James Bell Jr. is listed in exactly the same way: "James Bell Jr., Fishcurer, Peterhead."

Fishcurer?

I must admit that I am, well, shocked. The work sounds menial and brings to mind someone, his hands in brine, gutting a fish. Was that what the father of Eric Temple Bell did for a living before he emigrated to America? How can I tell Taine and Janet—and Lyle, who thought he would go to London and just look up the Bell Shipping Company in the telephone book—that E. T. Bell's father was, not the owner of a fleet of sailing vessels, but someone who cured fish.

There is more unexpected information on the marriage certificate of James Bell Jr., 40, and Helen Lyall, 25. Although it is the first marriage for Helen ("Spinster"), her bridegroom is described as "Widower." Here is one more fact that, to my knowledge, nobody in the Bell family knows. Who had been the first wife? Had there been children of that marriage? If so, had Bell known about them and perhaps even grown up with them?

Helen's residence is given on the marriage certificate as Nethermill, which lies in the parish of Cruden to the south of the parish of Peterhead. I learn, again through the 1902 edition of the *Encyclopædia Britannica*—that in the nineteenth century enough water flowed through Cruden parish to provide power for seven mills. Threadmaking was an important local industry. Could Helen Lyall as a young woman—and perhaps even as a child—have worked in the mills?

James Bell Jr.'s address is given on the marriage certificate as 24 Jamaica Street and on all the birth certificates as 16 Jamaica Street. The addresses are useful information, for with them I can find James Jr. and Helen in the Peterhead census of 1881, the first census taken after their marriage (although before the birth of E. T. Bell). There I may very well find the answer to my question regarding children of the first marriage.

I have no problem locating the family of James Bell Jr.

James Bell, Fishcurer, Head

Helen J. L. L., Wife

Enid L., Daur.

James R., Son

Also listed are a nineteen-year-old visitor from Cruden, a cook, and a nurse. But there are no children in the household other than the ones I already know about.

The perusal of the 1881 census gives me a vivid sense of what the town of Peterhead must have been like when Bell was born. Listed in the census are ropemakers and sailmakers, sparmakers, engine fitters, net menders, shipmasters and ship owners, fishermen by the score (some specifically identifying themselves as fishing off Greenland for whale and seal), fish gutters, sometimes herring gutters. There are shipbuilders, fish sellers and buyers, harbor and pilot masters, even a woman "deriving her income from sand." I am surprised at the large number of fishcurers. I note that fourteen of them indicate in their census listings the number of men, women, and boys that they employ. James Bell, however, does not list any employees, whether by choice or because he has none. He is simply James Bell, Fishcurer.

I learn something more about the general trade of Peterhead again from the 1902 *Encyclopædia Britannica*, where, it happens, the statistics given are for 1883, the year in which E. T. Bell was born.

The herring fishing, in which the port has long held a leading position (631 boats in 1883), was begun in 1818 by a joint-stock company. The general trade is of considerable importance. The chief exports are herrings (£180,000 in 1883), granite, cattle, and agricultural produce. In 1883 the number of vessels that entered the port with cargoes and in ballast was 864 of 87,839 tons, the number that cleared 840 of 86,318 tons. The town possesses ship and boat building yards, saw-mills, an iron-foundry, cooperages, agricultural implement works, woollen manufactures, breweries, and a distillery. In the neighborhood there are extensive granite and polishing works.

I cannot find a listing for James Bell in the preceding census of 1871. In the 1861 census, however, I stumble over him: *Bell, James, 24, Fishcurer, Foreman,* living as a boarder on Ship Row. I note that his landlady had two daughters.

By the time I receive the marriage and birth certificates for the Bell family from Edinburgh, I am on the eve of leaving for Peterhead, by way of London and Aberdeen. In the course of the trip I hope to be able to place E. T. Bell in the familial, educational, social, and economic background of his youth as well as to answer certain specific questions. I make a list of these, but I have learned by now that the answer to any question about Bell almost always leads to another question.

The weekend before I leave for London, I go down to Watsonville again, this time to tell Taine and Janet that Bell's father was not the owner of a fleet of sailing vessels, nor the owner of Penybryn, as Bell had led them to believe, but in fact a fishcurer in Peterhead, a little town on a rocky peninsula on the North Sea. It happens that my visit coincides with the family celebration of Janet's sixty-fifth birthday and her retirement from the practice of pediatrics. All the children—Lyle, Laurie, and Lisa—are there. I cannot say how they feel about the information I give them, which must be disappointing family history, but never at any time does any member of the Bell family suggest that I should cease what I am doing.

I spend much of the afternoon by myself in a corner quietly re-examining Bell's magpie memorabilia. I puzzle over books there that belonged to someone named Ellen Temple, and note again that the name Temple appears printed and in ink on the title page of Bell's father's *Caesar.* I find several books inscribed "To Ellen" from people whose names mean nothing to me. But there is one—Tennyson's *In Memoriam*—inscribed "To James from Ellen" with the date February 23, 1864, and a verse from the poem:

> Behold, we know not anything;
> I can but trust that good shall fall
> At last—far off—at last—to all,
> And every winter change to Spring.

Since February 23 is also the date of the inscription in a book of poems later presented "To James from Helen J. L. L. B.," I decide that it must have been Bell's father's birthday.

Although going again over the memorabilia does not seem too productive at the time, when I get home I come up with several new questions to be investigated on my trip. One, as I write to Taine and Janet, "is such a far out guess that I am almost embarrassed to mention it, but I am wondering if Ellen Temple may not have been the first wife of James Jr."

I have no one else to propose. It is at least worth checking. But *Eric Temple Bell*. How strange if James Jr. had given his second son by his second wife the maiden name of his first wife!

1861 – 1965

Extract of an entry in a REGISTER of BIRTHS

Registration of Births, Deaths and Marriages (Scotland) Act 1965

074631

No.	1. Name and surname	2. When and where born	3. Sex	4. Name, surname, and rank or profession of father, Name, and maiden surname of mother Date and place of marriage	5. Signature and qualification of informant, and residence, if out of the house in which the birth occurred	6. When and where registered and signature of register
47	Alice Dunphie Bell	1883, February Seventh 4h 0m Bell Petrshead	F	William James Bell Junr Fisherman Alice Dunphie Bell M.S. Lyall 16 January 1877 October 18 in Crimond	James W. Bell father	1883, February 20 At Petrshead Geo Robertson *Registrar*

The above particulars are extracted from a Register of Births for the .... District .... of .... Petrshead ....

in the .... County .... of .... Aberdeen ....

Given under the Seal of the General Register Office, New Register House, Edinburgh, on .... 31st January .... 1989.

The above particulars incorporate any subsequent corrections or amendments to the original entry made with the authority of the Registrar General.

This extract is valid only if it has been authenticated by the seal of the General Register Office. If the particulars in the relevant entry in the statutory register have been reproduced by photography, xerography or some other similar process the seal must have been impressed after the reproduction has been made. The General Register Office will authenticate only those reproductions which have been produced by that office.

Warning

It is an offence under section 53(3) of the Registration of Births, Deaths and Marriages (Scotland) Act 1965 for any person to pass as genuine any copy or reproduction of this extract which has not been made by the General Register Office and authenticated by the Seal of that Office.

Any person who falsifies or forges any of the particulars on this extract or knowingly uses, gives or sends as genuine any false or forged extract is liable to prosecution under section 53(1) of the said Act.

RX86/C 976

The Fishcurers of Peterhead

On September 7, 1989, my husband and I take a flight to Aberdeen, I through the generosity of the Vaughn Foundation and he through the courtesy of United's Mileage Plus Program. We spend the night in Aberdeen and the next morning rent a small car and drive thirty miles "north-northeast" to Peterhead.

I am eager to see the house where E. T. Bell was born. I realize that he did not live there more than a few years, perhaps no more than three, but his family on both sides was well established in the town, and from stories he later told he seems to have spent at least some school holidays in Scotland with his mother's family. In preparation for my visit, through the Arbuthnot Museum in Peterhead, I have sent local newspapers requests for information about Bells or Lyalls. I plan also to contact by telephone any families with those names still living in the town. Perhaps I will find someone whose parents or grandparents knew the family of E. T. Bell. I am also very interested in learning something about the nature of the business that Bell's father abandoned in order to grow fruit trees in the Santa Clara Valley.

Peterhead lies near the mouth of the River Ugie and was originally known as Peterugie, but its name was changed in 1560 when the land on which it stood was acquired by the Earl Marischals of Scotland. Later George Keith, the Fifth Earl, who founded Marischal College, came additionally into possession of the lands and livings of the Abbey of Deer. Although the Earl gave nothing from the Abbey estate to his new college, he was believed to have done so and was much criticized for the supposed misappropriation. In response he ordered the motto that Bell later used at the beginning of *Men of Mathematics* to be displayed on the buildings of his college as well as on various buildings in Peterhead. The motto, gone now, could still be seen during Bell's school days:

They haif said: quhat say thay: lat thame say.

In the Town Square, where a statue of the Fifth Earl Marischal stands, we find Scotland's ever useful Tourist Information Office. We obtain a map of the town, quickly locate Jamaica Street, where James Bell Jr. and his family lived, and proceed there. The street is only a couple of blocks long, sloping gently down to the protected harbor around which the town originally developed. At the time of Bell's birth a Masonic Lodge, which has since burned, blocked the present view of the water. Near the Lodge were hot and cold springs that in a still earlier day, before the development of the whaling and herring industries, had made Peterhead a popular spa. All this I learn reading the pamphlets we have picked up at the Tourist Information Office.

Jamaica Street is very clean, scrubbed by wind and rain, I think. It is surprisingly wide for a street in an old section of an old town and has generous stone walks on either side. The houses are constructed of large rectangular blocks of gray granite. From what we have observed, driving through the town, this gray granite construction has been common until recently, so the houses on Jamaica Street—well over a century old—do not seem antiquated. But in front of them there are no yards and no roses, as in Aberdeen. The houses are flush to the walks and flush to one another. One has the impression of a walled street.

Although it is a warm and sunny day, we are very much aware that we are on the coast of the North Sea. We imagine what it must be like in winter in those austere houses. An elderly woman comes out between a couple of parked cars and crosses briskly to a grocery store. There is no one else to be seen.

It is clear that the Bells' house at No. 16 was originally a single dwelling, but it has now been divided into two apartments. The apartment that is entered through the original front door is occupied by a Mr. Cameron, who agrees to show it to American visitors. Like others on Jamaica Street, the house is built parallel to the street so that it is wider than it is deep. We enter directly into the center of the house and go quickly through it to the yard in back. Mr. Cameron tells us that there are two floors and an attic. The stairway is narrow, probably because the building has been divided. Ceilings are low. There is, however, a deep yard in back where James Bell Jr. if he were inclined to horticulture would have had space to pursue his hobby.

We take our leave of Mr. Cameron. There is still time for a visit to the Arbuthnot Museum, which like other public buildings and fine homes in Peterhead is constructed of the famous pink granite of the area, said to be similar to that of ancient Egypt. The Arbuthnot Museum specializes in the history of the fishing industry and displays many implements of that trade. Photographs, some of them combined into a video presentation, provide a vivid glimpse of the curing yards during the day of Bell's father. Since it is Saturday, the director, Dr. Jocelyn Chamberlain-Mole (a skilled markswoman who competes in small-bore rifle matches) is out shooting, but I learn from an attendant that the Museum has received no responses at all to my newspaper queries about Bells and Lyalls.

The next day (Sunday) we drive to Cruden to see the church where Bell's parents were married. Nethermill, where his mother lived at that time, does not appear on modern maps of the area, but I am able to locate it on the century-old map that hangs on the wall of the Tourist Information Center. We are told that a couple of the other mills are still standing—one of them has been converted into a "lounge bar."

On our way to the church—now known as the Old Cruden Church —we stop to visit the most vaunted "place of interest" in the area—the rosy granite ruin of Slains Castle, which was visited in 1778 by Boswell and Johnson and a century later by Bram Stoker, who used it as the model for the castle of Count Dracula.

The church we are looking for lies among gentle hills broken by ancient burial mounds, but although quite near to "Dracula's castle" it cannot be seen from there. The spire that we do see is that of the Episcopal Church of St. James, which is even older than the Old Cruden Church. It has a font said to have come from the church erected by Malcolm II on the battlefield of Cruden to mark his victory over the Danes. In fact, we are told, the name Cruden derives from "crojudane," or "slaughter of the Danes."

The Church of St. James is open all day on Sunday, but the Old Cruden Church is locked, so all we gain from our trip to Cruden are bits of local lore and color. That evening I call all the Bells and Lyalls listed in the Peterhead telephone book, but none has any knowledge of the families in whom I am interested.

The next morning my husband and I present ourselves at the Arbuthnot Museum, where we meet Dr. David Bertie, the assistant curator. He has already located for me the three existing Peterhead town directories. In all three there is a listing for a James Bell on either Keith

Inch or Greenhill, the islands that once gave Peterhead its famous protected harbor:

1853   James Bell and Co., Fishcurers, Keith Inch
1864   James Bell and Sons, Fishcurers, 5 Castle St., 16 Greenhill
1877   James Bell, Fishcurer, Keith Inch

Since James Bell Jr. was listed in the 1861 census as "Foreman" and in the 1881 census simply as "James Bell, Fishcurer," I conclude that he was initially in business with his father and later established his own business.

I am interested in learning from Dr. Bertie about the status of a fishcurer in Peterhead.

"Was a fishcurer inferior to a fish factor?"

"Oh no! Not at all. A fishcurer was a person of considerable importance in the community."

He cites as an indication of social eminence the fact that between 1870–1880 four presidents of the Burns Society were fishcurers. James Bell Jr. was president from 1877 to 1879.

"It is a very great honor to be president of the Peterhead Burns Society," I am informed later by the present secretary. "It is one of the six oldest clubs in the world."

In Peterhead I have not forgotten my suggestion to Taine and Janet that Ellen Temple may have been James Bell Jr.'s first wife. While I examine the records of the Peterhead cemetery in the Arbuthnot Museum, my husband volunteers to search the *Peterhead Sentinel and Buchan Observer* in the local library, which is on the floor below the Museum, for a record of James Jr.'s marriage.

"Where should I begin?"

At the time my answer seems to me completely random.

"Try 1865."

"Why 1865?"

Why indeed? James could have married for the first time in any year after 1861, when he was "batching" in a boarding house, and before 1877, when he married Helen. Later I realize that I must have been basing my choice on the line inscribed by Ellen to James on his birthday in 1864:

And every winter change to Spring.

Happily for my husband the issues of the *Sentinel/Observer* during the first half of 1865 make very interesting reading. The American Civil War was coming to an end. Savannah fell. Lee surrendered. On an inside page a small item reported that the American president had been shot and was not expected to survive. When I come down from the Museum, my own search of the cemetery records having yielded nothing about James Jr.'s first wife, my husband is still reading but he turns back to a place he has marked.

"I was so interested in the war, I almost missed it completely."

I look where he is pointing:

On Thursday, the 4th of May, at All Saints Church, Upper Norwood, Surrey, by the Rev. G. Rockford Gull, B.A., James, eldest son of James Bell, Esq., of Upper Norwood, to Ellen, youngest daughter of the late Frederick Temple, Esq., of Guildhall, London, and Norwood, Surrey.

I cannot believe what is there before my eyes.

Ellen Temple and James Bell.

So James Bell had actually done it. He had given his son the maiden name of his deceased first wife!

That afternoon, drawn back to Jamaica Street, we go into the grocery store, which is on the side opposite from the Bell house, and ask the grocer about the history of the street. He is a handsome freshfaced and dark-haired Scot. He has not been on Jamaica Street long, but he tells us it is still a street inhabited almost entirely by men and women connected with the fishing industry. He is a former fisherman who now makes his living stocking the fishing boats. He tells us that we should have got up early that morning and gone down to the fish market.

"There is a market on Monday morning?"

In the old days the Scots never fished on Sunday.

"Oh yes, the biggest market of the week!"

Since he can't answer our questions about the history of the street, he passes us on to an elderly customer. She leads us to Mr. John Reid at No. 15 Jamaica, letting herself in without a knock. The fact that we also are named Reid starts the conversation. Mr. Reid, who is seventy-two, was a fishcurer, as were his father and brothers. When the curing industry died out in Peterhead, he went to work for British Petroleum. Now he is retired.

He willingly leaves his potting shed on a sunny afternoon—we are fortunate in our weather in Peterhead—to talk about fishcuring. He describes how, moving at lightning speed, the gutters—young women who worked at a piece rate—would deposit the herring guts in the "gut coggies," or small tubs that had been placed in big bins called "farlans," and then—without ever turning—would toss the cleaned herring into the appropriate "selection coggies" arranged in order behind them, often having to make as many as five different selections in that way. He explains in loving detail the curing itself: the packing, one layer of herring placed precisely across the layer below; the successive "filling-ups" to make good the sinkage that occurred as the herring settled; the insertion of the plug and the tapping by one of the "quines" with an ear expert enough to detect any small spaces remaining empty. He tells how the barrels were then inspected by the Fishery Officer, each barrel marked with the selection it contained and the name of the curer, the brine randomly tested by the inspector before the barrel was finally stamped with the respected "Crown Brand."

Mr. Reid concludes his discourse on fishcuring by leading the way upstairs to show us a book that contains pictures of the old sailing "zoomers" and "fifies." We eagerly examine the photographs in his book, for they are actual photographs of the kind of curing yards that Bell's father would have operated. In a central position there is always the long uncovered bin (or "farlan") into which wicker baskets ("creels") of herring are emptied. From the photos we estimate that at least a couple of dozen people could work comfortably around one farlan, although that many are never pictured. In one photo a cooper, wearing a bowler hat, scatters salt over the herring while another cooper, also wearing a bowler hat, turns them to mix salt and fish evenly. The young gutting "quines," wearing ankle-length oilskin "quites," work in teams of three, two gutting and one packing. We are told that these young women traveled around the country, following the catch. They could be seen in the evening strolling through the streets of Peterhead, knitting all the while, joking and chatting. Also to be seen were the tall Highlanders who came down to the coast for the herring season.

Mrs. Reid serves us tea with cookies, shortbread, freshly baked scones, butter and jam. We exclaim over the amount.

"*Just* tea," she says scornfully. "That's the English way."

The curing of herring in which Bell's father was active is long gone from Peterhead. The great "yards" on the islands of Keith Inch and Greenhill no longer exist, and the two islands are now one—and part of the mainland. The fishing boats still bring back herring; but in 1989, which is when we are there, the fish are going directly to black East

German boats lying in the harbor, and these are transporting them to large Russian ships, to be seen still farther out, against the horizon.

As I listen to Mr. John Reid, who is exactly the age of Dr. Taine Bell, I cannot help thinking that if James Jr. had not gone to America, Taine might also be puttering in his garden on Jamaica Street and describing to some stranger the great hauls of silver fish being brought from the trawlers into the yards, the seagulls hovering overhead, "so thick that the sky was white with their wings."

SEVEN GENERATIONS OF THE BELL FAMILY

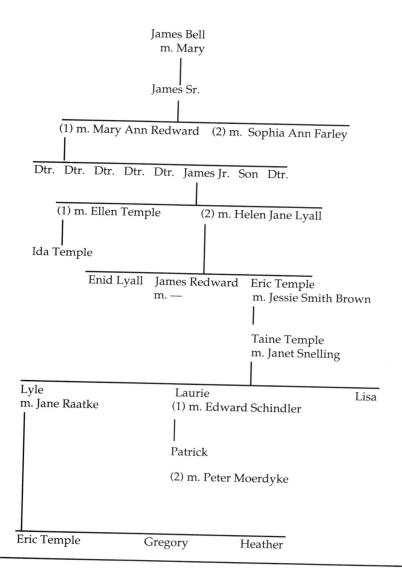

James Bell
m. Mary

James Sr.

(1) m. Mary Ann Redward   (2) m. Sophia Ann Farley

Dtr. Dtr. Dtr. Dtr. Dtr. James Jr. Son Dtr.

(1) m. Ellen Temple   (2) m. Helen Jane Lyall

Ida Temple

Enid Lyall   James Redward   Eric Temple
              m. —   m. Jessie Smith Brown

Taine Temple
m. Janet Snelling

Lyle   Laurie   Lisa
m. Jane Raatke   (1) m. Edward Schindler

Patrick

(2) m. Peter Moerdyke

Eric Temple   Gregory   Heather

From the Corner of Dark House Lane

In England I hope to discover three things: why Bell's father emigrated to America, the whereabouts of Helen Bell and her three children between the time of his emigration and the time of his death, and at least some understanding of the inexplicably secret role that his father seems to have played in the life of E. T. Bell. What follows is a chronological account of the English history of the Bell family, which I compile while I am in London in the hope of answering these questions.

For the purposes of my account, the progenitor of the family on Bell's father's side is the James Bell who was to be Bell's grandfather, James Bell Sr. He was born to a still earlier James Bell and his wife, Mary, on November 15, 1806, in the parish of St. Mary-at-Hill, which lies in the City of London a short distance up Fish Street from the Thames. At that time old London Bridge, a stone bridge that had been begun in the twelfth century, still stood and was still topped by two- and three-story buildings. The area around the bridge was a rough and disorderly one of fishermen, fishmongers, fish merchants, fish salesmen, and fish factors. Cockfighting was a popular entertainment and the strong, loud women who served as porters and sometimes as dealers had already made the name of Billingsgate synonymous with coarse, abusive language.

It was in this environment that the grandfather of E. T. Bell grew up. Probably he began to work in the Billingsgate Fish Market when he was quite young. The Market at that time was an open space next to the Billingsgate Dock, an inlet on the Thames River that, if it had been located on the coast, would have been called a harbor. In summer selling and buying began at three in the morning and in winter at five.

Salesmen and dealers protected themselves and their stock under large umbrellas and, when necessary, lighted their wares with candles.

"Here and in this way," a historian of the Billingsgate market has written, "huge quantities of perishable merchandise were sold and large fortunes realized."

When the James Bell of whom I am writing (later James Bell Sr.) was in his early twenties, he married Mary Ann Redward, a woman three years older than himself. By the time their first child was born, the couple were living in the Parish of St. Magnus the Martyr, which borders the river at London Bridge. Designed by Christopher Wren, the church has a lovely spire he intended to serve as a welcome to travelers coming up the Thames. Inside we again admire Wren's architecture and envy the purity of white and gold that would have surrounded James and Mary Ann when they came to worship. ("A mixture of incense and fish as well, I imagine," comments the Rev. Arthur T. Golightly, the sacristan, who generously takes time to tell us about the neighborhood during the period when James Bell was a fish factor at Billingsgate.)

Children came fast to James and Mary Ann Bell. Mary Ann in 1830. Maria in 1831. Lydia in 1834. Ellen in 1835. Adeline in 1836. Then, at last, on February 23, 1837, a son. He was named James, like his father and his grandfather, and as James Bell Jr. he was to be the father of Eric Temple Bell.

By the time James Bell Sr. became the father of a son, old London Bridge had been replaced by a new bridge (the London Bridge now in Arizona) and the neighborhood had improved. In an early directory of Billingsgate merchants I find a listing for James Sr. and his business on Lower Thames Street, which bordered the Market to the north. As a result I am able to find him and family on that street in the 1841 English census, presumably living above his place of business, as was the custom in the area.

The distinction among the various buyers and sellers of fish was not always clearcut at that time, I am told by the present Fishing Inspector of the Fishmongers' Company; but a fish factor such as James Sr. was essentially an agent, buying and selling fish for one or more principals. A later Billingsgate directory places him and his business precisely at No. 15, which is where Dark House Lane intersected Lower Thames Street. The lane, which no longer exists, took its name from a "dark house" on the corner that was known as a notorious rendezvous of fishwomen and seamen.

In spite of improved conditions around London Bridge, by the time James Jr. was six, James Sr. had established a residence in Norwood, one of the developing suburbs south of the City of London. That was where another daughter, Rosa, and a second son, Frederick William, were born. In the census of 1851 the Bells and their family of six daughters and two sons are listed as living on Central Hill in Norwood, which is the address that Bell's father, age thirteen, wrote in his Caesar. While the other children are described in the census as "Scholars, at Home," James Jr. is "At School."

With improved transportation, the result of the burgeoning suburbs, James Sr. was able to move easily by rail between Norwood and London Bridge. He foresaw future growth and demand for housing in the Norwood area and began to purchase long-term leases in the neighborhood, being partial to property in Upper Norwood.

Upper Norwood was an attractive retreat from the City with a healthful "country" climate that had appealed to the affluent since the time of Frederick the Great (who had come there during his final illness). In the 1850s, when Prince Albert's Crystal Palace Exhibition was dismantled and moved in its entirety from London to Upper Norwood, the area became even more popular, and a great many suburban "villas" (detached residences with gardens) were built on its many hills. Sometime between 1851 and 1861 James Sr. purchased or caused to have built one of these—the family home, Penybryn.

During this period the fish trade continued to be lucrative.

"It was ... more prosperous than now," the Billingsgate historian wrote regretfully in 1915. "The competition among all branches of the trade was less keen; profits were easily made."

The Market had been renovated. A roof had been added and a fine Italian facade built on the river side. By then James Sr. may have had two businesses on Lower Thames Street, for the 1850 directory of the Market lists a "Greenman & Bell" at No. 16 as well as "Bell, Fish Factor" at No. 15. At some point he became "Mr. Deputy Bell" and conducted his buying and selling from "1 Billingsgate," which was presumably the No. 1 stall in the Market.

He began to look beyond Billingsgate, and his eye lit upon Peterhead. By the 1850s Peterhead was the second largest herring port in Scotland with more than four hundred boats fishing for herring. Fishcuring had become very profitable, and the number of fishcurers there was rapidly increasing.

The fishcurer was the businessman of the booming herring indus-try. He "made engagements" to purchase at a set price the entire catch of certain boats, which were considered "his boats." (Could these have been "the fleet of sailing vessels" of which Bell told his wife?) He also oversaw the actual activity of curing in large "yards," which he rented and where he provided the necessary supplies and employees. He handled the inspection and the final sale of the herring, which was mostly on the Continent.

By 1853 "James Bell and Co." was established on the island of Keith Inch. In that year James Jr. was sixteen, and since boys of his class com-monly left school at fourteen it is quite likely that he was at Peterhead with his father. Since there are only the three existing Peterhead direc-tories, it is not possible to follow exactly the development of James Sr.'s business venture in Scotland, nor to ascertain to what extent he con-tinued to maintain his interest in his business, or businesses, in Lower Thames Street. His residence remained in the Norwood area, for in all censuses after 1861 he is listed at Penybryn.

The number of Peterhead fishcurers had grown from three in 1831 to thirty-six in 1861. In that year James Jr., twenty-four years old, was working as "Fishcurer, Foreman," according to the census. There is no census listing for his younger brother, Frederick William, but in 1864 "James Bell and Company" became "James Bell and Sons."

James Jr. was now in a position to take a wife. In May of 1865 he went down to Surrey and married Ellen Temple, who lived with her orphaned brothers and sisters on Crown Hill, Norwood, in what were described in censuses of the period as "the Grecian cottages." He had clearly known her from his boyhood in Norwood. The name Temple on the title page of his Caesar when he was thirteen could have been that of Ellen, who was two years younger than he, or that of her brother Josiah, who was a year or so older. At the time of the marriage of James and Ellen, Josiah Temple had recently become the head of the Temple family following the death of his only brother, Frederick George, a clerk in the Chamberlain's office in the City of London. As far as I have been able to determine, Josiah never married and there were never any progeny with the last name of Temple.

A daughter, Ida Temple Bell, was born to James and Ellen in 1866 but died at the beginning of 1867. Twelve months later, just before Christmas, Ellen also died. ("Consumption, two years," according to her death certificate.) She was twenty-eight.

At this point James Jr. disappears from the documentary record. I have no idea whether his absence is significant. He resurfaces in 1877, the year in which he married Helen Jane Lindsay Lyall. From that time on, his life is reasonably well documented. He and Helen had three children. They named their first son James Redward in accordance with the Scottish tradition that the first son is given the name of his father's father and the maiden name of his father's mother. By the same tradition the second son is given the name of his mother's father. Since, however, Helen's father was also a James, the second boy was named Eric and for some reason given the maiden name of his father's deceased first wife.

The year of Eric Temple Bell's birth was also the year in which his grandfather, James Sr., who had been widowed since 1879, decided to marry again. The lady of his choice was Sophia Ann Farley, a spinster, also residing in Upper Norwood, a few years younger than he. Since Sophia Ann's father, a wealthy wine merchant, had left her very well off, there was a pre-nuptial agreement between her and James Sr. that kept her fortune in her own hands. Nevertheless, making a new will, he left her all his personal property and a life estate in Penybryn.

My husband and I are able to examine James Sr.'s will at Somerset House, the British depository for all such records since the middle of the nineteenth century. The will is thirteen pages long with three codicils added during the next three years. Although the custom of primogeniture still prevailed in England, James Sr. was most concerned with providing for the women in his family and the motherless children of his daughter Rosa. He was generous, noting that the support he had already provided for Rosa's children was not to be set off against their portion of his estate; but he was also determined to prevent anything he left at his death from falling into the hands of the men his daughters had married. In all cases "powers of appointment" were specifically withheld. If the daughters predeceased their spouses or, in the children's case, their father, what they had received in the will was to revert to the estate. This determination on the part of James Sr. makes the disposition of Penybryn especially interesting. First, a life estate to his wife, Sophia Ann. Then "upon trust for my daughter in law Helen Jane Lindsay Bell the wife of my son James Bell."

Penybryn, the crown and symbol of his success, the only property he owned outright, was to go, not to his eldest son, but first as a life estate to his second wife and then as a life estate to "the wife of my son James." There was no other mention of James Jr. in the entire will.

He was to receive nothing. After Sophia Ann's death, Penybryn would go to Helen and her assigns for life, so that she could live in it or lease it for income, "and after her decease upon trust for all and every the children and child [sic] of my said daughter [sic] who shall be living at her decease and who shall have attained or thereafter live to attain the age of twenty-one years for their own absolute use and benefit as tenants in common."

"Our London home," as Bell called it, would go ultimately to him and his siblings and—perhaps simply by oversight—to any other children that Helen, then only thirty-one years old, might have in the future by James, or by some later husband.

My first reaction is that James had been purposely passed over by his father, "disinherited," since any claim of his as a pretermitted heir had been closed off by the mention of his name. But there is another possibility, my husband points out. From the text of the will it is clear that the younger son, Frederick William, had been buying James Sr.'s share of the business on Lower Thames Street with money lent to him (at interest) by his father. Upon James Sr.'s death, however, whatever debt remained was to be forgiven and Frederick William was to take the business free and clear. That was his father's sole bequest to him. Possibly James Jr. had had a similar arrangement with James Sr. in regard to the Peterhead business and had already paid up in full. This conjecture is supported by the fact that in 1877, when James Jr. married the second time, he was listed in the Peterhead directory individually and not, as in earlier directories, in business with his father.

In the British Newspaper Library in Collinwood I search through the Peterhead newspapers to see if there is an item about James Bell Jr.'s leaving for America. I order the issues from 1883 (when I know that he was present at his younger son's birth) up to 1886 (when I know that he was in San Jose) and quickly find myself absorbed in the unfolding story of what was happening during that period in the herring industry in Peterhead. The catches had been exceptionally good in the years immediately following the birth of Eric. They were so good in fact that they shortly resulted in the glutting of the European market, on which about seventy-five percent of the Scottish herring industry depended. Even before the 1886 season began, it was clear that there was going to be a serious depression in the industry. The depression especially affected the fishcurers, since their business required advancing large amounts of borrowed money to purchase supplies and to finance the "engagements" with their boats. In 1885 there was an entire column devoted to the bankruptcies of curers. The following year the herring industry was in even worse shape.

The curers refused to make any engagements in advance. They would purchase the catch only at auction after the boats came in.

The fishermen went on strike, the hired Highlanders and lively joking "quines" went home. One boat after another was beached. I find nothing about James Bell Jr.'s leaving Peterhead, but I do find that in 1886—the same year that a listing for James Bell first appeared in the San Jose directories—the Peterhead newspaper began to run a column on its front page advertising cheap and easy emigration to America.

Is it possible that I have stumbled upon the reason that Bell's father left his business in Scotland and emigrated to America?

SANTA CLARA COUNTY AND ITS RESOURCES,
"a historical, descriptive, and statistical account," published by the *San Jose Daily Mercury* in 1896 and reissued by the San Jose Historical Museum Association in 1987

In Gold Coin of the United States

In Scotland and England I am able to learn quite a bit about Bell's family background but nothing at all about his boyhood—for the very good reason, I am becoming increasingly certain, that he was not in the British Isles for most of his first thirteen years.

As a result of examining James Sr.'s will I have been able to dispose of the doubts created by Lyle Bell's statement that "in 1888 or 1889 Penybryn passed from James to Helen." It is now clear that Lyle's statement was based on two natural but erroneous assumptions: (1) that "James Bell" was James Bell Jr. and (2) that "Mrs. James Bell" was James Jr.'s wife, Helen. In fact, she was James Sr.'s second wife, Sophia Ann Bell, who came into her life estate in Penybryn at James Sr.'s death in January 1888.

From various records I have seen in London I have been able to push back the dates when I know or can reasonably assume that Bell's mother was in San Jose: in 1896, when James Jr. died; in 1892, when she was identified in Sophia Ann's will as being of San Jose, California; in 1888, when she was not listed among the mourners at the funeral of James Sr. The significant fact, as Sherlock Holmes observed, is that the dog did not bark. The fact that I have found no evidence of the presence of her children in Scotland or England—other than their births—adds to my conviction that they were with their mother and father in America.

But I am still hoping that while I am in London I can establish with finality the presence of the three Bell children in San Jose. Before I left home, I wrote to the British Society of Genealogists to inquire if records of passengers from foreign countries entering English ports in

1896 still exist. The director of the Society responded promptly that if there were such records they would probably be among the Home Office Records at the Public Record Office in Kew "though some of them may not yet be available to public search under the 100-year rule." The Search Department of the Public Record Office responded with equal promptness: "The PRO holds Board of Trade inward passenger lists for the years 1878–1960 . . . ." That was the good news. The bad news was "but there are only a handful surviving for 1878–1899."

Nevertheless, the morning my husband and I set out for Kew, I am hopeful if not optimistic.

The branch of the Public Record Office in Kew is an amazingly efficient modern building, quite different from the ancient edifice in Chancery Lane where earlier we have seen the Domesday Book, the first English census. Upon presenting our passports, we each receive a "reader's ticket" and, for 20 pence, two lead pencils saying "Public Record Office." We are directed upstairs to a room where we are given beepers that will inform us when the materials we have ordered are available. We are then sent to another large room containing card catalogues and reference volumes. Under BT 26 (Board of Trade) we select six volumes—we are each allowed to order three—covering inward passenger lists from January to May 1896 at three English ports. I have asked the clerk at the desk which ports are closest to London but, being a child of the age of flight, she has no idea of travel by passenger ship. We select London, Southampton, and Liverpool on the basis of our reading of English literature. We order these by number on computers and then go down to the cafeteria for lunch. We are eating an English stew when both our beepers go off.

Up at the desk we are told that the boxes of materials we have ordered are so large that we will have to examine them in the map room. There, on a high table surrounded by tall stools, we find six big boxes. They contain the original passenger lists on long and often rolled up sheets, yellowed and literally cracked with age. My husband chooses to take the London lists, a poor choice it turns out since most of the ships arriving there come from India. He finishes his boxes while I am still checking Southampton, where most of the arriving ships are out of New York. I let him have Liverpool, which has been our third choice. I have just finished Southampton when he calls me over.

And there they are, in the perfectly clear, round hand of the purser of the *S.S. Britannic* out of New York—the returning Bell family, second cabin passengers, arriving in England on April 10, 1896, four months after the death of the husband and father:

Helen Bell	42
Enid	17
James	15
Ernest [*sic*]	13

At last—incontrovertible proof that Eric Temple Bell was in America *before* he came to Stanford University in 1902.

But for how long? As so often in research, as in mathematics, the answer to one question raises another.

The last day we are in London we take a train to Norwood to seek out the grave of James Bell Jr. From a lengthy obituary that appeared in the *Norwood News* I have learned that James Bell Sr. was buried in the Norwood Memorial Park, and I assume that Helen Bell would have returned her husband to be buried with his father and other members of the Bell family. I hope that the cemetery records will give the cause of his death in America. Of course they do not—by this time I should know that nothing affecting James Bell Jr. will come easy. However, the cemetery records do tell me that he was interred, not in the family plot, but in a plot he purchased in 1867 when his infant daughter died. He lies with her and her mother, Ellen Temple Bell.

Given a detailed map of the cemetery at the desk, my husband and I wander back to the old section to find the grave. It is overgrown with weeds, and there is no stone on it. We are able to identify it only in relation to the grave of James Sr., which is two plots above it on a slight rise. For James Sr. there is still a tall stone bearing his name and the names of the "beloved first wife" and "the beloved loving second wife" as well as those of his younger son and one of his daughters, each with an appropriate line from scripture. There is a line from *Ecclesiastes* under the name of James Sr. and, recalling his will, we agree that it is appropriate: *Though dead, he speaketh.*

Returning our map at the desk, we ask the clerk if the inscriptions in the cemetery were ever copied. She says that she is certain that they were and gives us the number of the Minet Library to call. The librarian there is regretful. Yes, the inscriptions were copied, but by then the stone in which we are interested was already gone.

On the flight home, drafting a progress report for the Vaughn Foundation, I write to James Vaughn Jr. that as a result of information I have obtained on my trip much of the material about Bell's early life and family background presently in print will have to be revised. Bell was born, not in the university city of Aberdeen (*Who's Who* and others) but in the small fishing town of Peterhead. His father was not "a merchant mariner" (*National Cyclopedia of American Biography*) but a Peterhead fishcurer who emigrated to America and became an orchardist. His mother did not come from "a family of classical scholars for several generations back" (*Twentieth Century Authors*) but was the daughter of a Peterhead schoolmaster whose father was a farmer. Bell did not grow up in England (*Dictionary of Scientific Biography*) but spent at least most—in my progress report I guess at ten—of his first thirteen years in San Jose, California.

I am uneasy about making that last statement. I am convinced that I am right, but it *is* possible that Helen and her children could have joined James Jr. sometime after he established himself in San Jose. If I am going to write about Bell's life, I have to be able to state:

E. T. Bell came to the United States on such and such a date when he was so many years old.

To do so, I need to find out exactly when Bell's father purchased his American property. The day after our return to San Francisco, I suggest to my husband that instead of unpacking, putting things away, and going through the accumulated mail, we pretend that we are still in London. We will not take the Underground to the Public Record Office or the British Newspaper Library, but instead we will drive down Highway 280 to San Jose, a distance of perhaps fifty miles from San Francisco, and do our research in the Santa Clara County Recorder's Office. He thinks that is a great idea, and off we go at eight o'clock in the morning.

Transfers of property in the early days in San Jose were recorded, as they are today, alphabetically and chronologically according to Grantor (seller) and Grantee (buyer), each in a separate index. Thus when James Bell's property was sold after his death to John A. Fair, the transfer appeared both in the "B" index of Grantors and in the "F" index of Grantees for the year.

We begin our search with the microfilm for Grantees in 1883 since, although I know from Bell's birth certificate that his father was present in Peterhead at the time of the birth, it is possible—if not probable—that he left for America that same year. But there is no Bell among the Grantees of 1883, or 1884, or 1885, or even 1886 when *Bell, James* was first listed in the San Jose City Directory as the owner of property there. We continue through the next decade in case the recordation of the deed was delayed, as it sometimes is. In all those years there is no James Bell listed as a Grantee. There are other Bells, and we ask to see the deeds of the various properties that they purchased on the chance that Bell's first name may have been incorrectly recorded. But none of these properties matches the description of Bell's property, which we have by metes and bounds from the description for the estate sale.

In the coming months we make a half dozen trips to San Jose and propose many theories to explain how a man who owned a piece of property at his death would never have been recorded as having purchased it. Ultimately we are able to chart the transfers of the property from the time of its subdivision as two separate parcels in the early 1870s up until the time both parcels came into the hands of one Frank Garst, who purchased them on August 1, 1881, for $4000. In six months, pocketing a profit of $225, he sold the property to John A. Angus. A year and a half later, on June 12, 1884, Angus sold it to a newcomer from England named James Bell "for the sum of Seven Thousand and Five Hundred ($7500) Dollars gold coin of the United States of America to him in hand paid."

June 12, 1884. The date is unexpectedly early—and significant, for it establishes that James Bell Jr. did not emigrate to America because of the depression in the herring industry, which first began to be felt in 1886. He left Peterhead and came to America while the herring industry was *booming.*

Before we say goodbye to the patient and helpful clerks in the Recorder's Office, we go back and recheck the Grantor and Grantee indexes for 1884. We find that, although the record of the sale of the property by Angus to Bell appears in the "A" index of Grantors, there is no corresponding record of the purchase by Bell from Angus in the "B" index of Grantees. This is odd, since for the legal record of the ownership of property it is more important to know who purchased the property than to know who sold it. If I were inclined to believe in the supernatural, I would think that there was somebody that did not very much like what we were doing.

"I still cannot believe," my husband says, "that of all the transfers of real property in Santa Clara County in 1884 we would be looking for probably the only one for which the Grantee was never listed in the Grantee Index!"

But now, at last, I have an upper bound (June 12, 1884) as well as a lower bound (the date of Bell's birth in 1883) for James Bell Jr.'s emigration to America. I hope only that his family came with him; otherwise

I have no idea how to trace them. But to what port would they have come? Not to San Francisco, I am certain, for by then people were no longer sailing around the Horn or crossing by train at the Isthmus of Panama and continuing their voyage up the Pacific Coast. I think New York is the most probable port of arrival, but there is always the possibility of their coming to Boston or to one of the numerous ports south of New York—or even to Montréal, which was Bell's own port of entry when he came back to America in 1902. By now my husband and I have had enough experience going through passenger lists in England to know that if we have to search a twelve-month period, we can do so for only one port. It *has* to be the port of New York.

From the National Archives in Washington D.C. I order microfilms of the New York passenger lists from June 1, 1883, to June 1, 1884. Six weeks later eleven rolls arrive at the Federal Records Center in our area. The following Saturday we are at the entrance when the Center opens. On the drive down we have decided not to try to "psych out" James Jr. but to proceed chronologically through the rolls of microfilm. I take the first six.

By lunchtime we have each come across several "Bells" and have both lost a little of our confidence by almost passing over one or two of them, lulled by the progression of names, Irish, English, Scotch, Scandinavian, German, Italian, Polish. The experience of the morning has thoroughly convinced us, if we were not convinced before, that unless the family came together and came to New York there is not any way we can hope to ascertain when E. T. Bell arrived in the United States.

After lunch we move faster than in the morning when we have been slowed down by the migration unfolding before our eyes. At 3:45, when the warning buzzer rings for the Saturday closing, we are both on our last rolls. Then, just a few minutes later, my husband says, "Come and look at this."

I bend down over his shoulder and see on the machine:

James Bell, 47, British citizen.

The date of arrival is May 22, 1884, less than a month before the purchase of the property in San Jose. I do a quick calculation.

"The age looks right," I concede.

In a way I am more disappointed than excited. So, I think, he came alone. Then I see the names below his.

Quickly, because the Center is closing and there is not time to make a copy, I note the details of the ship: the *S.S. Circassia* out of

Glasgow. The master, one Archibald Campbell. A stop on the way at Moville, on the north Irish coast, to pick up more emigrants. A total of 638 passengers, of whom 560 were traveling in steerage. There were sixty-one passengers in second cabin, and seventeen in first cabin. Of the seventeen, eleven were returning American citizens. Five of the other six first cabin passengers were the Bells. They had come well heeled to the new world, and they had come to stay:

James Bell, 47
Mrs. Bell, 31
Enid L., 4
James R., 3
Eric T., 1 [yr.] 3 mos.

The old rascal! He came to the United States when he was fifteen months old—probably before he could even walk. He lived there, in San Jose, California, for twelve of his first thirteen years. And he never told anybody.

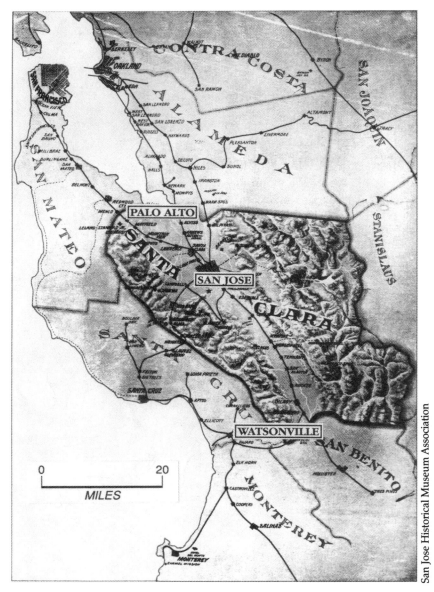

A CALIFORNIA VALLEY "not over-broad . . . a man might ride in a day"

San Jose Historical Museum Association

A California Valley

The Bell family disembarked in New York on May 22, 1884. Their journey by train across the American continent would have required four days and three nights, taking them through country very different from any they had known. They would have arrived in Oakland, California, which was where the railroad ended. There they could have taken the ferry across the bay to San Francisco, if they had wished to see that already fabled city, or they could have gone directly by another train to San Jose, which lies near the southern end of San Francisco Bay.

San Jose was then as now the principal city of the Santa Clara Valley. It had a population considerably larger than that of Peterhead as well as three daily newspapers, seven banking institutions, the largest fruit cannery in the world. There was a "world class hotel" where wealthy foreigners and easterners came to winter in the California sun, and there were a number of substantial residences, spectacular conglomerations of frame with porches, bow windows, towers, and cupolas. Public buildings and other important edifices were of the local sandstone, which was of a rosy hue vaguely similar to that of the famous Peterhead granite. At the center of the town was a 237-foot tower topped by a 15-foot flagpole. At night, strung from bottom to top with electric lights in a series of circles of ever decreasing size, it resembled a gigantic Christmas tree and illuminated the entire "downtown."

The Bells were arriving too late to see the entire valley in bloom, as it would have been at the beginning of the year, but acres and acres of trees were already heavy with fruit beginning to ripen. From the promptness with which James Bell proceeded to San Jose and purchased property there, it can be assumed that before he left Peterhead he had obtained a considerable amount of information about the advantages, economic and climatic, of the Santa Clara Valley through the kind of promotional material that was being disseminated at the time. San Jose "boosters" were only too willing to explain the fortuitous factors responsible for their valley's phenomenal fertility. They were also eager to back up their claims with figures: the price of land, the price of young trees, the cost of pruning, spraying, irrigating, fertilizing, drying, sacking, selling.

Although I can find no item in the San Jose newspapers about the Bell family's arrival in town, I do find an advertisement running in the *Daily Mercury* from the end of May to the beginning of June that may very well describe the property that James Bell purchased so soon after his arrival:

> $7500 17 acres about 2 mi from town fine two story house
> good barn windmill etc 6 acres in orchard good soil.

I learn from *Sunshine, Fruit and Flowers,* a book published by the *Daily Mercury* in the 1890s and since republished by the San Jose Historical Museum Association, that the varieties of fruit most satisfactory and profitable in Santa Clara County at that time were prunes, peaches, apricots (which grew in few other localities), and cherries. Prune growing was the most extensive and profitable fruit industry, and well over a third of the trees in almost every orchard were prune.* The trees commenced bearing in the fourth year, sometimes even in the third, and being dried the fruit did not have to be marketed immediately. Next most popular were apricots, which could be sold fresh, or dried in the sun, and which were becoming more popular all the time. Bell's novels are full of references to apricot trees in bud or bloom, and it is fairly certain that his father's orchard consisted largely of apricot and prune trees. (I guess at prune because the orchard James Bell purchased was obviously a very young one.) There were a number of fruit packing, canning, and drying companies in the area; and James Bell could not have failed to note that although there were trees and ladders instead of the fishermen and their boats, fruit boxes instead of creels and barrels, drying and canning instead of salting, his new occupation bore a certain similarity to fishcuring.

---

* Although the fruit is in fact a plum that dries without fermenting, the word prune was indiscriminately used at the time for the tree and its fruit as well as for the dried product.

The preceding paragraphs would have constituted the extent of my knowledge of Bell's life in the Santa Clara Valley before his father's death in 1896 if once again he had not given himself away. Glancing over a list of unpublished manuscripts donated by Taine and Janet to the University of California at Santa Cruz, I am startled to see the title "A California Valley." This is the only title on the list with which I am not familiar. Is it possible that Bell actually *wrote* about his boyhood?

A few days later I have the manuscript in hand.

It is the typescript of a fairly long poem, 567 lines, headed as below, but with everything except "A California Valley" crossed out:

> Out of Doors
> A California Valley
> To W. F. H. and K. A. B,
> Wherever They Now Are.

From dates on other poems in the same folder, it appears that "A California Valley" was written in 1916. That was twenty years after the thirteen-year-old Bell left the Valley following the death of his father. Although he never speaks in the first person in the poem and never identifies the valley of which he writes, his poem is an ode—a veritable hymn in praise of the Santa Clara Valley. Unfortunately very little of it can be said to be specific to his boyhood experience in the Valley. More incidents, although still not a great many, come from his later life there as a student at Stanford. But the creek described in the following lines appears on the survey map of his father's property (which the poet always refers to as "a ranch"), and its course can be followed even today in the modern subdivision: "and behold! the later noon/ Ripens the fruits of gold, and slackens a creek that flows/ All year round the orchard's garden to a sluggish, emerald pool."

There are other lines as well that describe the orchard:

> Mighty fig-trees towering by windmill or barn or well,
>> Strew their bountiful riches purply over the ground
>>> Where the revelling sparrows chatter all summer, and deign to share
>>> Their feast with the rancher's children, and other birds more rare . . .
> Perhaps a lost Mexican parrot who swears like an infidel.
>> Then a dropping of fruit in the bucket makes a most musical sound
> And the sun-dried apricots, bleaching, exhale a summery smell.

But Bell does slip in one very sharp sketch of a little boy at play on "the ranch," and this little boy can only be the little Eric Bell:

Who dwells in those tiny caves on the sunniest bank of the creek?
"Lizards?" the small boy asks,—with a stick and patient research;
"A horned-toad!" he gleefully shouts, as out with a rolling lurch
Scurries the monstrous dwarf of an antediluvian freak—
(One of those scaly beasts who lunched on the village church),
Old Ceratops, sadly diminished, and friendly, the last of his line.

One naturally wonders about Bell's schooling in San Jose and that of the other children. In *Twentieth Century Authors* he wrote, "My education was by tutors until I entered the Bedford Modern School." It is possible that his mother, a schoolmaster's daughter, could have taught her children at home, or that a young tutor could have been obtained inexpensively. But the Bell children may also have attended one of the many schools in the area. Santa Clara County considered itself an exceptional educational center. In addition to fourteen private schools, there were eighty-six public school districts, one hundred and fifty-four primary schools, one hundred and four grammar schools (two of which were near the Bells' home), and four high schools.

In addition, in 1892, when Bell was nine, the county became the location of one of "the finest and most richly endowed" universities in the United States. If the Bell family ever went up to Palo Alto, a few miles north of San Jose, they would have seen the handsome buildings of Santa Clara County sandstone that still form the "Quad" of Leland Stanford Junior University. The new university had been founded by Senator and Mrs. Stanford in memory of their only son, who had died at fifteen, and it was committed to the belief that "A generous education is the birthright of every man and woman in America." It was unusual—perhaps unique?—in that it charged no tuition.

"Such munificence is unparalleled in history!" exclaimed a contemporary writer in the *Daily Mercury*.

The one thing that can be said with complete certainty about Bell's education in San Jose is that he and his brother and sister, like most children in the community, would have been excused from lessons whenever fruit was ready to be picked. From the list of agricultural equipment in James Bell's estate, it is clear that his orchard was a one-man operation, but the presence of eight ladders in the estate indicates that family and hired hands were put to work when the time came. In later life Bell often referred to his experiences picking fruit in the Santa Clara Valley, but the implication was always that he had worked in the orchards during vacations at Stanford in order to pay his living expenses while a student there, never that he had picked fruit as a boy.

After I finish reading "A California Valley," the loving lines tumbling out, there is no question in my mind that the dozen years Bell spent as a boy in the Santa Clara Valley were for him as close to paradise as he was ever to come.

Twenty years later it was the beginning of the year when "fresh from their brief sweet slumber . . . the orchards awake" that he was to recall in loving detail:

A carmine swelling of buds,—and behold! a blossomy snow
    Whitens the apricot orchards over a night, and bees
    Inaudible yesterday, sing through the honeying trees
Till the warm low winds of the Valley hum with the music they make;
    Another day and a night, and warmer the breezes blow—
    On the orchards a shimmer of pink is the miracle morning sees:
Long miles of peach-trees blossom, with apricots alternate;
    Then all burst out in bloom, plum trees, apples and pears—
        No paltry few by a farmhouse,—broad counties of lesser lands
        Would melt in these flowery seas . . .

It was in the first days of 1896, just when the acres of almond trees were coming into bloom, that this early and idyllic period in Bell's life came to an end. At that time he was just a month short of thirteen. New Year's Day fell on Wednesday. On Saturday, apparently around eight o'clock in the evening (a deduction based on Bell's series of "In Mem's" in his father's Caesar), James Bell Jr. died. It was the anniversary of this death that was to be so frequently and cryptically observed by Bell throughout his life.

There are several curious circumstances surrounding James Bell Jr.'s death. As I mentioned earlier, the death was not recorded by Dr. Pierce, who was presumably the family doctor as well as a trusted friend. For a death, even by natural causes, at which a doctor was not present, the Coroner was required to hold an inquest. There is no record of an inquest into the death of a James Bell in the Santa Clara County Coroner's Book, which appears to be well kept and complete. In the notice of the death and the funeral that appeared in two of the three San Jose papers no cause of death was given. It was stated in the obituaries that interment would be in Surrey, but in the Clerk's Probate Register, which includes all bills paid by the Estate, there is no record of embalming or shipping expenses. The only bill paid, other than $10 to the East-Side Fruit Growers Union, is $1000 to one William Y. Miller. In the local directory for that year Miller is listed simply as "a Reverend," but earlier directories place him at the Presbyterian Church in East San Jose, which is where the Bells lived. (Helen Bell was a Presbyterian.) It could only have been Miller who handled the return of James Bell's body to England. The fee for this service was considerable—an eighth of the gross value of the estate—one might say exorbitant when one considers that even ten years later it was possible to cross the United States by rail for less than thirty dollars.

These circumstances are so puzzling to me that I decide to write to Judge Mark E. Thomas Jr., now retired from the local bench and an authority on the legal history of Santa Clara County.

In his opinion, I ask, is there some reasonable explanation, in the circumstances I describe, for the absence of records in the death of James Bell?

A few days later, the Judge responds by phone.

"I have given quite a bit of thought to your question, and I am really at a loss. If the death had occurred a few years later I would say that the circumstances you describe would have been most unusual. It would have looked very suspicious—as if something were being covered up. But even at the earlier date in which you are interested, it is odd. There is certainly enough there to arouse one's curiosity."

Helen Bell informed her family in Peterhead of her husband's death by telegraph. An obituary notice, which also did not give the cause of death, appeared on January 14 in the *Peterhead Sentinel and Buchan Observer*, a copy of which I am furnished by that newspaper. I receive as well a surprisingly long article from the same issue about James Bell Jr.—surprising in that he had been gone from the town for more than a decade and yet the article about his death was longer than was customary except in the case of distinguished citizens.

In Peterhead he had been "well and favorably known as an intelligent and honorable trader." While there he had taken "a lively interest in the affairs of the town [and] evinced much intelligent interest in the matters connected with the harbours. Although he had never been a member "of any of the public boards he [had] never ceased to interest himself in all that pertained to the welfare of the town." This interest in the town's welfare had continued after his departure for California where he had also "carried on a successful business."

It is questionable how successful James Bell's business in California had been. He and his family had come to America first-class. With cash in hand he had purchased what was clearly a very young orchard for $7,500, and almost a dozen years later it was appraised in his estate at only $8,000. In the interim there had been one of the most severe financial crises of the nineteenth century—the "Panic of 1893." (There is a reference to this disaster in Bell's unpublished novel *To Be Kept*, in which the father of the gifted hero-scientist "loses everything" in "a panic" but dies "free of debt.") The Notice of Sale of Real and Personal Property of James Bell listed in addition to his real property, the following personal property:

> Fives shares stock in East Side Fruit-Growers Union, one spring wagon, one set of team harness, one spray pump, one lot fruit boxes, one plow, one harrow, one-horse cultivator, one clod smasher, one fruit truck, eight fruit ladders, one light bedroom set, one dark bedroom set, one black horse.

There were no debts other than the ten dollars owed to the Fruit Growers Association.

Helen Bell could have chosen to remain in San Jose and manage the orchard at Lucretia Avenue and Story Road, the children's half of their father's estate being held in trust. Women orchardists, farmers, and landowners were not uncommon in the Santa Clara Valley. But she had other plans. Even before her husband's estate was settled, she set out across the continent with her three adolescent children.

A formative chapter in the life of Eric Temple Bell closed, not to be reopened even to his son. But the California Valley of his youth was not to be forgotten: "For he who has lived there a year is homesick the rest of his days . . . ."

TWO

a habit of independent work

A Maker of Mathematicians

The destination of Helen Bell, disembarking at Liverpool with her three fatherless children, was Bedford, a busy market town some fifty miles from London and equidistant from Oxford and Cambridge. She had good reasons for settling there.

As the *Encyclopædia Britannica* was to note a few years later, "in proportion to its size, Bedford has more public endowments than any other place in the kingdom, for which it is chiefly indebted to Sir W. Harper, Lord Mayor of London in 1561, who founded here a free school, and conveyed for its support, and for portioning poor maidens, a piece of ground in London, the surplus, if any, to be given to the poor." The Lord Mayor's land had risen in value over the centuries to such an extent that by the time Helen Bell and her family arrived in Bedford, it supported five schools: grammar, modern, and preparatory schools for boys and two high schools for girls, all of which were virtually free to students. Such inexpensive education attracted many families, most often retired civil servants and military people. These usually left the town as soon as their children had completed their schooling, and were disparaged by the permanent residents as "squatters."

Helen Bell proceeded to Bedford with the same dispatch with which she and her husband had earlier proceeded to San Jose. Between April 10, when the family arrived in Liverpool, and the beginning of May, when the new school term began, she had settled in a modest two-family house on Foster Hill Road and had enrolled her sons in school, James at the Bedford Modern School and Eric at Orkney House, a small preparatory school. At seventeen Enid was too old for admission to either of the girls' high schools.

I do not know why Helen Bell placed her sons in different schools. Andrew Underwood, the archivist at the Bedford Modern School, does not think the fact is significant.

"Perhaps there was no place at the Modern. Many families in town had one boy at the Grammar School and one at the Modern."

But unlike those two schools, Orkney House wasn't a Trust school, so Helen Bell would have had to pay tuition for her bright younger son. She may have hoped that he would profit from a school like Orkney House, the primary goal of which was to prepare its boys for entrance examinations at the great English public schools—and, in the case of the more gifted, for scholarships.

It is difficult to conceive the "culture shock" which thirteen-year-old Eric Bell must have experienced upon entering Orkney House. He had spent virtually all his life in the Santa Clara Valley of California and had never known anything other than the American West. He came to Orkney House and an undemocratic, "class" oriented environment, not as an English boy or a young Scot, but as an *American.*

Orkney House no longer exists. The information about the school that follows comes from several nineteenth century pamphlets describing its faculty and offerings.

The founder and master of Orkney House was A. H. Blake, an "old boy" of the Bedford Grammar School, who in his time had garnered one of its Exhibitions (i.e., scholarships) to Cambridge and had studied classics there. He was assisted by four "form" masters in addition to visiting masters for French, Music, Carpentering, Drilling, and Gymnastics. The forms (classes) were small, averaging only a dozen boys, so Bell would have received a considerable amount of individual attention if not "the education by private tutors" to which he later referred. The subjects taught were Latin, Mathematics, English, History, Grammar, Geography, Divinity, Spelling, Dictation, French, and German. (Greek is also listed in one of the pamphlets I examine, but not in others.) Mrs. Blake endeavored to make the boys look upon Orkney House "more as a home than as a school." Perhaps this was another reason that Helen Bell selected it for her younger son.

When thirteen-year-old Eric entered Orkney House, he was at the age that often marks the beginning of an interest in mathematics. It is natural to wonder whether there was any indication of mathematical ability or interest in his family. On one occasion later in life he described his father as a good mathematician and his maternal grandfather (the parish schoolmaster in Peterhead) as having trained several good mathematicians. It is also perhaps relevant that the one book that he inscribed as passing directly from his father to himself on the day of his father's death was Charles Kingsley's romantic novel about Hypatia, the daughter of Theon the Mathematician and reputedly the first to edit the *Arithmetica* of Diophantus—the source of the theory of numbers, the subject to which Bell was to devote his professional career. Although Diophantus is not among the mathematicians mentioned in the novel (Euclid and Apollonius are), it does not seem improbable that the imagination of a thirteen-year-old boy would have been captured by the story of a beautiful woman who lectured on mathematics and was cruelly murdered in Alexandria by a mob of Christian fanatics.*

The year 1896, which had begun for the Bell family with a death, was also to end with a death. On December 18, 1896, Sophia Ann, the widow of James Bell Sr., was buried with her husband in the Norwood Memorial Park. She is described on his gravestone as the "beloved loving second wife," and the adjective *loving* seems to have been well chosen, for she remembered each of his daughters in her will and also remembered the wives of his sons. At the time she wrote the will (1892) James Jr. was still alive in California but his younger brother, Frederick William, had died, leaving a widow and small children. Sophia Ann seems to have considered James Jr.'s wife in San Jose as bereft of a husband as the widow of Frederick William, for she divided the income from a thousand shares of stock between them, the stock itself to go to the survivor. With Sophia Ann's death, Helen Bell also came into her life estate in Penybryn with the sole and exclusive right to its possession and its income until her death.

Many years later, in the autobiographical notes he wrote for *Twentieth Century Authors,* Bell emphasized almost more than any other information about himself the fact that "at one time" he had lived "on a hill overlooking the old Crystal Palace grounds." The Crystal Palace of which he wrote was a remarkable structure of slender iron rods and plates of clear glass that had formed the great hall of the Prince Consort's innovative exhibition in London in 1851. When the building, which had been constructed of prefabricated parts, was moved to Upper Norwood a few years later, one of the exhibitions that came

---

* In Bell's final book, *The Last Problem,* he titled his chapter on Alexandrian mathematics "From Euclid to Hypatia."

with it was a collection of "life-sized" models of dinosaurs.* The scientific consultant for their construction had been Richard Owen of Cambridge University, who gave dinosaurs their name, which means "terrible lizards." The only time that Bell could have lived at Penybryn would have been during a school recess when Helen Bell went there to remove Sophia Ann's things and prepare the house for lease. During that very brief stay the sight of the "great amoral brutes," as he called them in *Twentieth Century Authors,* not surprisingly marked the beginning of what was to be "a life-long fascination" for the young teenager who a few years earlier had been collecting lizards and horned toads in the Santa Clara Valley.

Back in Bedford at the beginning of 1897, now with increased income from stocks and rents, Helen Bell moved her family to a single-family "villa" on Clarendon Road. The houses in the new neighborhood all had names. The Bells' home, a yellow brick building with a carved lintel above the front door, was *Hillsborough.*

Bell remained at Orkney House for two years, but the only documentation of his time there is the book among his memorabilia inscribed "Orkney House, First Prize" and dated April 1897. At the beginning of January 1898 he entered the Bedford Modern School, where his older brother, James, was still in attendance.

The two Trust schools for boys—the Modern School and the Grammar School—were both held in high regard in Bedford, I am informed by Mr. Underwood. The Modern School, however, began at a later date than the Grammar School and in an inferior position. Originally known as the Writing School, it was set up to take boys from eight to fourteen, before they began their apprenticeships in the various trades, and teach them to read and write. At a much later time the distinction between the Modern School and the Grammar School was made as follows by a local newspaper, according to Mr. Underwood's history:

"The Grammar School prepares for those professions in which a classical education is the chief requirement. The Modern School trains for those professions in which the classics hold a place of subordinate importance, but in which what are generally called 'modern' subjects are essential."

---

* Although the Crystal Palace itself burned to the ground in 1936, the original dinosaur models are still displayed in a natural setting within walking distance of the Crystal Palace station in Upper Norwood.

THE AUTHOR at Bell's home with Neil Gallagher of the Modern School

Bell arrived at the Bedford Modern School in the last years of the stewardship of its famous head, Dr. Robert Poole. "The Doctor" had attended Rugby and, although by his time the great headmaster of Rugby, Thomas Arnold, had passed on, the Doctor had imbibed the Arnold ideal of what an English public school should be. According to

Mr. Underwood, critics accused him of trying to give his school a status it was never intended to have, but nevertheless he succeeded in bringing the Modern School almost to the level of the Grammar School while always paying lip service, as in mathematics, "to the applications."

During the Doctor's last years, which coincided with Bell's two years at the Modern School, age and ill health had resulted in a diminution of his driving energy. The School's physical plant had deteriorated to "an appalling state," according to a report by the school inspectors. In many classrooms the light, both natural and artificial, was insufficient. The unplastered walls afforded "ready lodgement for dust and dirt." There was no central heating system, only a few coal-burning stoves, and the buildings were "miserably and unhealthily cold." The seating was bad. "There is hardly a room in which the boys can reach the ground with their feet when sitting on the forms [in this case, benches]. The seats are also far too narrow, while no forms have backs." For all this information I am again indebted to Mr. Underwood's history of the Modern School.

During Bell's first year he was addressed, and so designated on the school rolls, as "Bell minor" since his older brother was still in attendance.* From the evidence of their school courses the Bell brothers were quite different in interests and abilities. "Bell major" was enrolled in "the Commercial Side." This was a course of study held over from the days when the Modern was much less academic. The boys who followed it were generally looked down upon, and for this reason the course was shortly to be abandoned. "Bell minor" was enrolled in the Higher Modern Side, the most academic course offered, very close to what he would have had at the Grammar School except for the lack of Greek. Boys who followed the Higher Modern Side were expected to go to the University or to Woolwich (the Royal Navy's training college) or into the India Service. In getting into the University, however, they were handicapped by having had fewer years of preparation, the compulsory leaving age at the Modern School being seventeen while at the Grammar School it was eighteen.

James Bell was also apparently somewhat bigger and better coordinated than his younger brother and more sports oriented. Proficiency in sports was of great importance for "a good public school man." Years later, in *Desmond,* another unpublished novel, Bell was to

---

* In the family circle James was "Hippie" and sometimes "Flapps" while Eric was "Hoggie."

rail against "the endless grind of futile games" at public schools and to describe his hero as "[having to submit] to what he considered the senseless discipline of a leisured class which has nothing less boresome to do with its time."

On Bell's first afternoon at the Modern School there occurred what was to be one of the most significant events of his professional career. As "a new boy in the Upper Fifth Form," he had his first contact with Edward Mann Langley, who would make a mathematician of him.

Bell would later recall it vividly when he wrote of Mr. Langley.

"Almost before he had entered the room, the powerfully built man with the auburn beard had begun. It must have been evident even to those who had no particular liking for mathematics that they were lis-

Mathematical Gazette

EDWARD MANN LANGLEY: "What he sacrificed his pupils have gained."

tening to a mathematician. To those whose previous training in mathematics had been at the hands of classical scholars, the lesson was a revelation."

Among the boys Mr. Langley was known as "Bogus" for that word with which he dismissed all mathematics that he did not consider genuine. Bell later recalled the occasion early on in Mr. Langley's class when he first brought himself to the teacher's attention. The assigned problem involved what mathematicians sometimes describe as "dirty mathematics." The other students gave up when they could not obtain the kind of clearcut answer they expected, but Bell successfully slugged his way through.

Although Langley was the highest paid master at the Modern, his salary was only £250, or $1250, a year. The Doctor had suggested at various times, according to Mr. Underwood's history, that Langley apply for higher paid posts and had even arranged an offer for him from one of the great public schools, but Langley was attached to the town and to the School, which he himself had attended, and preferred to stay. It was Bell's good fortune.

When I visit the Bedford Modern School during my trip to England, I am taken by Mr. Underwood to his office, where he has laid out copies of the *Eagle,* the School's magazine, and a small leather-bound book that contains a handwritten record of the various forms during Bell's years at the School. Next door a mathematics class is in session, and the steady murmur of the boys' reciting their lessons accompanies my researches.

The form record details exactly Bell's academic progress. He entered the Upper Fifth Form in January 1898. At Easter he was listed as seventeenth in a class of twenty-nine, but by Midsummer he was second in a class of thirty.

He was then promoted to "Sixth Form-Remove." This is an intermediate stage, according to Mr. Underwood, a boy being promoted from Remove to the Sixth Form on the basis of fitness in age and character to be in the highest form. At the beginning of January 1899 Bell was promoted to the Sixth Form. By Easter he was eighteenth out of twenty-five on the Headmaster's special list; at Midsummer, tenth out of thirty; at Christmas, fourth out of thirteen. The following Easter he was second out of twenty-four; and at Midsummer, when he had to leave because he was seventeen, he was first on the Headmaster's list.

"Well, he did make it," says Mr. Underwood, "didn't he now?"

The courses Bell was taking during his last term were listed by letters: *e, f, g, l, m, s*. English, French, German, Latin, Mathematics. I ask Mr. Underwood if the *s* stood for "science," which had been introduced during Doctor Poole's tenure.

"No. Scripture, I should think."

The Modern School had a longtime affiliation with New College at Oxford, and its "leaving boys" took an examination under the auspices of the College that was equivalent to what is today the Advanced ("A") Level Examination. For successfully passing this examination, Bell received a "Higher Certificate" from Oxford University. He and six of his classmates also took the Matriculation (or Entrance) Examination for the University of London. He finished in "First Class" with two other boys, but neither he nor any of his classmates received "Honours," although on occasion he sometimes listed himself as having done so.

Bell's progress through the Modern School had been accompanied by various prizes in the form of books. These are still among his memorabilia in Watsonville: Midsummer 1898, a prize for English; Midsummer 1899, a prize in political economics; Midsummer 1900, prizes for scripture, for general subjects, and for physical laboratory work as well as the Form Prize. Oddly enough he received no prize for mathematics; yet it was mathematics that he took with him from the Modern School.

"Whatever success I have had as a mathematician I attribute entirely to Mr. Langley's coaching," he later wrote to the boys of the Modern School. "No ordinary good teacher can impart the knack of research even to a quick-witted student. The ability to do scientific research probably is largely inborn. Nevertheless a mind with the capacity to do research usually needs a flick at the critical moment by some similarly constituted mind, to start it off in the right direction. By a great stroke of good luck I fell in with Mr. Langley at the most impressionable period—when I was about fifteen. He administered the necessary flick—as painlessly as was possible under the circumstances. Being a mathematician himself, as well as an extraordinary teacher, he knew how to create *mathematicians*, not mere examination-passers."

E. T. BELL at eighteen after leaving the Modern School

Going Their Separate Ways

In the year 2001 the English census of 1901 will be made available to the public. At that time it will be possible to ascertain what, if any, employment Bell found for himself between leaving the Modern School at seventeen and returning to America at nineteen. Until then, the principal sources of information about his life between 1900 and 1902 will have to be two documents: the obituary of Mr. Langley that he wrote in 1933 for the *Mathematical Gazette*, founded by Langley, and a letter he had written earlier to Mr. Langley at the request of the editors of the Modern School's magazine.

The letter is a valuable source of information, not only about Bell's education in England and the United States, but also about his professional career up until 1932. Its writing came about in the following way, according to the editor of the *Eagle* of the Bedford Modern School (also founded by Mr. Langley):

"A rumour having repeatedly come to the Editor's ears that one of the greatest—if not the greatest of all—mathematicians in America was an O.B.M. [Old Bedford Modernian], he very shrewdly applied to Mr. E. M. Langley for information. Mr. Langley—kindly as ever—promised to write to this distinguished Old Boy ... and ask him to send the *Eagle* an account of his career. The latter speedily replied in the form of a letter to his old teacher, who has given us permission to reprint the greater part of it."

Published in two installments as "The Career of Dr. E. T. Bell (O.B.M.) in America," Bell's letter described, not only his own mathematical career, but also "the necessary steps by which a young man in the United States enters on a mathematical career." Subsequent quotations in this chapter and subsequent chapters, unless otherwise attributed, come from this letter.

In 1900, when the seventeen-year-old Bell had to leave the Bedford Modern School, his formal education in England was at an end. But his education did not end.

Private lessons in mathematics with Mr. Langley continued, or perhaps began—it is not quite clear from the manner in which Bell refers to them—but the nature of the lessons is clear from what he wrote in his obituary of Langley:

"In private lessons he gave prospective pupils one disciplinary instruction, and it was sufficient. An unprepared assignment, except for reasons of health or accident was automatically to sever relations between master and pupil.

"[Mr. Langley] 'made haste slowly' at first, insisting that the pupil find his own way, even if it took a week. Then when the pupil had begun to acquire the habit of independent thinking, Mr. Langley would rapidly outline the next topic, usually in writing, and give the pupil two or three days to complete a rather heavy assignment on it. He demanded clear expression, and practically all the work was written out in final form before being shown to him for criticism and correction."

Since Bell was not preparing for university or civil service examinations, Mr. Langley let himself go and took the young man through analytical geometry, elementary mechanics, and the calculus, with lots of algebra and geometrical conics (his own favorite subject) as well. Seeing that Bell was not fond of geometry, he suggested "a look at the beginnings of the theory of numbers" and illustrated with "an ingenious use of Wilson's theorem [that gave] a very elegant proof of the determination of the quadratic character of two."

Mr. Langley awakened Bell's interest in another subject to which the younger man was to devote study—elliptic functions. He gave no formal instruction, "but awakened my interest by showing how some rather abstruse identities concerning homogeneous products followed at once from elliptic identities. This had come up, in some way which I cannot recall, in connection with a theorem due to Clifford, for whom Mr. Langley had a limitless admiration as a geometer."

Some thirty years later, after Langley's death, Bell still remembered "more than one ghastly private session with Mr. Langley, in which precisely nothing tangible in the way of problems was done during the whole hour, simply because he insisted that I see for myself the things which any rational human being of my age should be expected to see." At the time, he confessed, he did not very much enjoy such sessions. "I did not realize that what seemed like no progress at all was really the only progress that mattered."

During these two years Bell was also being privately tutored in Greek. The earliest ambition of his life had been "to know the Greeks at firsthand through their literature." He had attempted to learn the language on his own but had not got very far. Now Greek had been added to the curriculum at the Modern and, although Bell was no longer a student there, the new Greek master, C. L. Ryley, agreed to take him as a student. Ryley's method of teaching was most effective, Bell reported.

"Even to-day I can read Greek easily after only a day or two spent in reviewing the verbs. 'The surge and thunder of the Odyssey' has stayed with me all these years, and it has been one of the great things I got out of my early education."

On several occasions in later years Bell attempted to account at least partially for the gap in his curriculum vitae between the Modern School and Stanford University by stating that he attended the University of London from 1900 to 1901, or in one case from 1900 to 1902. On most occasions, however, he wrote accurately that he "matriculated at the University of London but did not attend."

During this two-year period, from what he later told young women friends, he fell in love for the first time. This first love, he told them, was always the great love in one's life, and no other would ever equal it. I know nothing about the particular young woman involved, but a similar statement occurs in his novel *Desmond.* The heroine there is the daughter of the vicar of "Old Wootin Church," which is described as lying "nestled among its famous elms." In Bell's memorabilia there are two snapshots of *Old Worden,* a village owned by the Duke of Bedford. He identified one as "the Avenue of Elm Trees near Old Worden" and noted that it gave "a very poor idea of the actual place." There is no one in the picture.

During this period, between Bell's leaving the Modern School and his returning to America, his older siblings were coming into their inheritances from their father's estate (about $850 apiece at the time of his death). Enid had reached her majority in August 1899, and James had reached his in November 1901. Bell himself would not be twenty-one until February 1904.

Although I have continued to be curious about these siblings, I have not learned much about them or why Bell endeavored, all his life, to prevent any contact between them and his son. True, I have learned a little about Enid's future life from letters Bell wrote to Taine during and after the Second World War. By that time Enid, who had never married, was living on the Isle of Wight, where she died in 1959. As to the future life of James Redward Bell, supposedly in Africa, I have learned nothing.

When, however, Lyle was in England in 1981, he had looked up the will of Enid at Somerset House and had learned that she had distributed her estate among thirteen women friends and relatives and a Rear Admiral Edwin Howard Drayson, her executor. I decide to send letters to all Enid's legatees in the hope that I may find someone who knew the Bell family in England. One by one my letters come back. *Deceased. Unknown. Moved. Try Scotland.* (This last comes from Australia.) Then, nine months later, when I have given up all hope, I receive a letter from Jill Drayson Morgan—the daughter of Admiral Drayson, the executor. She writes that Enid Bell lived for many years in the family of her grandparents, her role being similar to that of a maiden relative. The reason that she has delayed writing to me, Mrs. Morgan explains, is that some members of her own family think that she should not answer my letter since the only details she can give me are "hearsay family stories" about the Bell family, "not very complimentary to people long dead."

"[Enid Bell] was a very beautiful woman, but very embittered. The story goes that her father, James Bell, was cruel to his wife and children. . . . As regards James Redward Bell, he was reputed to be a bad lot. He died in Ceylon in 1947. My sister [Vivian Drayson] was with Enid when she got the news, and I am sorry to report that she [Enid] thought it was a good thing."

Not Africa at all, but Ceylon!

Shortly after I receive Mrs. Morgan's letter, I am able to obtain from a directory of "Old Bedford Modernians" the Ceylon address of James Redward Bell: Meria Cotta, Meskaliya. At almost the same time I come across an article in the *Economist* on the continuing existence of slavery in Asia, Africa, and Latin America in the form of bonded laborers "[who] work in conditions often indistinguishable from slavery." Modern Ceylon, or Sri Lanka, is specifically cited in the article. It occurs to me that James Redward Bell might have become a plantation-owner or perhaps the overseer of a plantation, and the "slaves" of the tale his brother told about him might well have been the Tamil labor described in the article as having been imported from southern India at the end of the nineteenth century to work on tea plantations in Ceylon. Meskaliya is, in fact, in the hill country of Sri Lanka, a short distance from Kandy—the center of the great tea-growing plantations.

In the course of reading about the island, I learn that it was once known as Serendib, from which comes the word *serendipity*. I also learn that if a foreigner can prove that he has a regular external income the Sri Lankan government will allow him the privilege of settling there. The longest resident westerner under this program, so I learn, is the science fiction writer Arthur C. Clarke.

What serendipity!

I write to the author of *2001* on the chance that he may have some interest in such predecessors in the genre as "John Taine." I hope only that he will be able to give me the name of some person who can do research for me—I expect to pay for the time and expense involved.

Ten days later, at bedtime in San Francisco, my telephone rings.

"This is Arthur Clarke," a booming voice announces. He has received my letter. "John Taine" was one of his heroes. He will be delighted to help me, but his producer is arriving the next day to put together some thirty television shows. It will be a month or so before he can get back to me, but I will be hearing from him. Before hanging up, he gives me his phone and fax numbers. I say, "Good night," and he says, "Good morning."

It is two months before we establish contact again—this time by fax. To my surprise Clarke takes on the research himself. At first things go slowly. Inquiries in the neighborhood of Meskaliya turn up nothing. The National Archives reports no information.

"Can you be reasonably certain of his date of death before I hire someone to hunt through the newspaper morgues?"

Before I can check the date with Vivien Drayson, who was with Enid when she received the news of James Redward Bell's death, a fax arrives from Clarke.

"EUREKA!"

He has sent requests for assistance to two of the Colombo newspapers, and they have run good-sized notices with headings that attract attention: *Does anyone know J. R. Bell?* in one paper and *Who knows Bell?* in the other.

There have been instant results. The next day W. A. Roberts, the husband of James Redward Bell's youngest child, calls on Clarke.

Rohan De Silva

ARTHUR C. CLARKE at his desk in Colombo, Sri Lanka

"He's a well-educated ex-navy man who speaks excellent English," Clarke reports. "However his wife, Beatrice [James Redward Bell's daughter], only speaks Sinhala so she couldn't answer any questions directly."

According to Mr. Roberts, James Redward Bell fathered four children by Muthusamy Rangamma, but Beatrice is his only surviving child. Although he never married the mother, he supported the family until his death.

One after another, people respond to Arthur C. Clarke's request for help—a government minister, a plantation agent, a former government archivist, planters. From what he writes of these various contacts I am able to piece together a rough picture of James Redward Bell's life. He returned to England during the First World War and served at the front as a courier for a cavalry regiment. In 1920–1921 he was listed in Ferguson's Ceylon Directory as "2nd Lt., General List, Needwood Group, Haldummulla." By 1925 he was established in Ferguson's as "Planter and Manager, Upper and Lower Ohiya Estate, Ohiya." That listing remained the same up through the 1934 Directory, the most recent available at George Steuart & Company, Ltd., which was the agent for the Ohiya Plantations. (From 1930 on he also served as a justice of

JAMES REDWARD BELL, second from right, with friends in Ceylon

the peace.) He returned to England during the Second World War, but apparently not for active service this time. He died in 1949.

A letter from the son of a neighboring planter gives me a more personal picture of Bell's brother and his life in Ceylon.

"I was only nine years old when I saw [J. R. Bell] for the first time in the year 1918. He was at that time Manager, Needwood Group, Haldummulla in the Uva district. Everytime I saw him he was on horseback, frequently riding in the company of two other European gentlemen from the neighboring estates. It was a regular feature that these three gentlemen were always together and spent most of their week-ends either fishing off Horton Plains or swimming in the southern coast of Hambantota.

"[Mr. J. R. Bell] was a well built person with a burly figure, always with a cigarette in his mouth, being a chain smoker. He was a popular figure among the planting community and had a wide circle of friends. He was known as 'JR' among his friends while the local people called him 'TIGER' for his temperamental moods at times."

Another descendant of J. R. Bell also responds to Clarke's newspaper query. He is a grandson, Cyril Richard Wilfred Bell, whom Clarke describes as "a very intelligent and well-educated man, a quality control inspector recently back from working in Saudi Arabia." He brings with him a portion of his father's application for British citizenship, the claim to citizenship being "by descent." Under "Particulars Relating to Applicant," the following information is provided:

Place of birth:	Ampitiya Udahentenna
Date of birth:	5th February 1904
Name of father:	James Redward Bell
Nationality of father:	British Subject
Name of mother:	Muthusamy Rangamma
Nationality of mother:	Tamil

It is clear from the date of the first child's birth that the relationship between James Redward Bell and Muthusamy Rangamma was no casual liaison. It began at least as early as his twenty-second year and lasted until his death.

I am not able to obtain from the daughter or the grandson any specific information about the relationship of James Redward Bell with his younger brother. They know that he had a brother in America and that the brother wrote books. But that is all.

Enid and James may have already gone their separate ways before Bell left England. In later years he sometimes said that he "ran away" to America; however, the few existing documents do not support the story of running away. Among his unpublished manuscripts there is a short poem titled "Goodbye." The last two verses are highly derivative, but the first verse contains some specific details, such as a Scottish proverb, which may give it authenticity. The mother is speaking:

> "Goodbye, or for a year farewell;
> Take one last look on me,
> "Then turn not back lest luck turn too.
> Remember the laburnum tree,
> "My hand on the gate, and yonder yew.
> One year, no more, goodbye!
> "Your home is here."

The Bible his mother gave him as a parting gift is among his memorabilia at Watsonville. It is inscribed "To Eric Temple Bell from His Mother H. J. L. L. B." and dated June 28, 1902.

At the beginning of July the nineteen-year-old Bell boarded the *S.S. Tunisian* at Liverpool. The ship was not of the best quality. It carried no first-class passengers and was slower than most ships crossing the Atlantic at the time. On July 13, 1902, according to passenger lists furnished by the National Archives of Canada, "E. T. Bell, Gent." disembarked at Montréal. Crossing Canada by rail, he entered the United States at Sumas, Washington, a small town north of Seattle (information obtained from his naturalization papers). From there he proceeded down the Pacific Coast to San Jose, California.

He never saw either of his siblings or his mother, or England, again.

E. T. BELL at twenty, a student at Stanford University

The Return to Eden

At the beginning of September 1902 Eric Temple Bell enrolled at Stanford University, which was then entering upon its twelfth year as a coeducational institution with free tuition. The University consisted of a number of low sandstone structures and a grassy quadrangle—"the Quad." These sat in the middle of the acres of rolling hills and live oak trees that had formed the ranch of Leland Stanford and his wife.

Thirty miles up the peninsula that sheltered the southern end of San Francisco Bay was the exciting, rowdy, romantic, always capitalized City. That was the place where rich students in tuxedos went dining and dancing with fashionably dressed young ladies while the less affluent contented themselves with wandering through the notorious Barbary Coast, talking and drinking in one of the City's innumerable saloons, or having dinner in some little basement Italian restaurant, a basket of sourdough bread and a bottle of red wine on every table. Twenty miles to the south of Stanford lay San Jose and the orchards of the Santa Clara Valley.

The fact that Stanford University charged no tuition was undoubtedly important to Bell, but more important emotionally was the fact that in coming to the young American university he was returning to the paradise of his boyhood. In his poem, "A California Valley," he later described a horseback ride made by a young man, alone and in the moonlight, to the "salty marshes" of the long southern arm of San Francisco Bay:

> By a sudden turn in the road all Eden breathes once more—
>   A Fragrance of laurel-leaves crushed, with lavender, mint and thyme,

Red apricots ripe in the sunset, and almonds abloom in the
breeze,
Magnolias dewy with dawn, and loquat flowers,—all these
Meet in one keen reminiscence of places known long before
Man was a restless exile cold in an alien clime—
For there in the moonbeams quiver ten golden mimosa trees!

The few specific incidents described in "A California Valley" are
not always easy to follow. Bell was no Tennyson, and the complicated
rhyme scheme he chose often forced him into awkwardnesses that obscure his narrative. Even here the telling words do not leap out at the
reader, but they are there: *all Eden breathes once more . . . in one keen reminiscence of places known long before man was a restless exile cold in an alien
clime.*

As far as I have been able to determine—and this is a qualification
I must always make—Bell never revealed anywhere but in this stanza
how he felt about being swept up from the warmth and beauty of the
Santa Clara Valley and set down, "a new boy," in a cold, gray English
town and a new and very different school.

In the fall of 1902, however, he was at last back in the Valley. Reg-
istering at Stanford University, he gave as his "Parent or Guardian" Dr.
R. E. Pierce, who had been the executor of his father's estate. As evidence of his educational qualifications he presented his London "Ma-
tric" and his Oxford "Higher Certificate." Such details come from his
1931 letter to Mr. Langley that was published in the *Eagle*. Again, all
the first person quotations in this chapter, unless otherwise attributed,
are from this letter.

"I applied for the chance to crowd what remained of the undergraduate course into two years. The authorities put me 'on probation':
if I passed the term examinations, I was to go on; if I failed, I was to be
set back."

Bell was confident of his ability to make the grade, for his own
background was not the average one in America.

"My great advantage was the flying start I got at the Modern under Mr. E. M. Langley, and the really exceptional training in independent work which he gave me in private lessons. When I entered Stanford University . . . , I had already acquired the one thing which is of
the very first importance for a research career—*the habit of independent
work.*"

His first semester at Stanford he registered for seven courses (eighteen hours), all in the Department of Pure Mathematics. The mathe-

Stanford University News Service

H. F. BLICHFELDT and G. A. MILLER of Stanford's mathematical faculty

matical faculty consisted of four men: two full professors, one associate professor, and one assistant professor. Neither of the full professors had a Ph.D. but, as Halsey Royden has pointed out in his "History of the Stanford Mathematics Department," Robert Allardice and Rufus Lott Green were established mathematicians who had been at the University since its beginning. The Ph.D.'s on the faculty were G. A. Miller and H. F. Blichfeldt, both of whom had studied abroad. Miller later went to the University of Illinois, where he married well, invested wisely, lived frugally, and left a fortune to the mathematics department of that university. Blichfeldt headed the Stanford mathematics department from 1927 to 1938.

Like Bell, Blichfeldt had emigrated as a youth, but from Denmark. He had then worked as a farm laborer, a millhand, and a surveyor for an engineering firm. The latter, recognizing his mathematical ability, had urged him to go to college. In 1894, at the age of twenty-one, he had been admitted to Stanford "on probation"—like Bell. Completing the university course in two years, he had gone to Leipzig for a year with Sophus Lie mastering the Lie theory of continuous groups.

According to the 1902–1903 Stanford directory there were twenty-nine undergraduate students in mathematics, four graduate students, and one special student—that would have been Bell. He signed up for Determinants (Blichfeldt), Advanced Co-ordinate Geometry (Green), Differential and Integral Calculus (Allardice), Differential Equations (Blichfeldt), Geometry of Three Dimensions (Allardice), and Theory of Equations (Green).

"The lectures (third year of the special course in mathematics) plunged me at once into *modern* mathematics. . . . To my surprise I discovered that these modern things are in fact *easier* than the old stuff, probably because they are less incoherent."

In addition to courses in mathematics proper, the future historian of mathematics and biographer of mathematicians elected a one-semester course in the History of Mathematics taught by Miller. Among Bell's books at Watsonville is *A Short Account of the History of Mathematics* by W. W. Rouse-Ball, which he purchased on September 10, 1902, undoubtedly for Miller's course. I am interested in finding out whether he had already developed an interest in the history of his subject, perhaps through Mr. Langley. But, although the latter was a man of many interests—among these English literature and botany (he was an authority on the hybridization of blackberries), I find no mention in his obituaries of an interest in the history of mathematics.

Grades in mathematics courses at Stanford were customarily *pass* or *fail* and indicated by plus or minus. Bell received pluses in all his courses and on January 5, 1903, was granted "full entrance standing and *sixty* hrs. toward graduation."

The second semester he took another course from Miller, this one on the theory of numbers, the subject to which he would devote his professional career; however, his real introduction to that subject came, not through Miller or any faculty member at Stanford, but through the German mathematician and writer on number theory Paul Bachmann. A full professor at Münster, Bachmann had retired at fifty-five to devote himself to what was ultimately a six-volume summary of contemporary number theory. There are five volumes by Bachmann among the books Bell later gave to start a little research library for mathematics at Caltech. He inscribed one of these, on the theory of quadratic forms, "(First reading 1902 at Stanford with Chaiskov, second reading at Seattle 1908)." But I can find no one named Chaiskov in the Stanford faculty/student directories. That semester Bell also assumed a heavier load in mathematics, adding the Theory of Functions (Allardice).

"The instruction at Stanford was a curious mixture of the French, German and American methods, but it was extremely effective. . . .

PAUL BACHMANN "spared no pains to make clear his reasons"

The professors insisted on a thorough mastery of general principles rather than skill in problem-solving. Their attitude was this: if a problem was worth the serious effort of an advanced student, it should have at least the elements of a research project in it, and not be merely a difficult puzzle whose solution would add nothing to existing mathematics. . . . The American influence was to be seen in the way the problems were done—at the blackboard, once a week, under the professor's vigilant eye, and sometimes, caustic tongue."

Although Stanford charged no tuition, Bell still had to buy books and pay for room and board. In his 1931 letter to the *Eagle,* he emphasized, *"Few American students have their entire way paid through the university by their parents, and fewer still by scholarships."* There were not even many scholarships in the United States in comparison with the number available in England. "Contrary to what seems to be the universal opinion of us abroad, young Americans do not have unlimited money to spend. A few do, of course, but in the States we try to give those without mere money a chance to make good. The result is that the majority earn at least part of the money required for their education; many earn every red cent."

At this point Bell might have revealed what he himself did to earn money, but instead he wrote, "It would doubtless astonish most English readers to scan even a partial list of the jobs an American student is not above tackling in order to pay his way through the university. All things considered, I think it is better to spare the reader's feelings and omit such a list." In his opinion, "having been English myself," the hardest thing for such a young man to master on coming to the United States "is the theory, current here, that no job is in itself menial, although many men, no matter what position they hold, are menials."

In future years Bell was always to write with a touch of bitterness about students who did not have to work while attending school. In his unpublished novel, *To Be Kept,* the hero, who has won a tuition-free scholarship at a university where rich men send their sons but has needed to work to pay his other expenses, is described as "scarred to the bone" by years of "fighting relentless poverty." And writing the biographical memoir of Blichfeldt for the National Academy of Sciences in 1945, Bell made the point that Blichfeldt "did not have to take odd jobs, as so many did, to pay his way."

There is some information about how Bell supported himself at Stanford in an interview by Lee Shippey that appeared in the *Los An-*

geles Times in 1933. He worked his way, Shippey reported, by tutoring students for examinations.

"If they failed to pass the examination they paid him nothing, but if they passed they paid him double. He was so successful that he not only made them pass but supported himself, and was able to support a couple of other Britishers."

I cannot say much about Bell's personal life during his days at Stanford. Certainly it was a new and (from what he later wrote) a pleasant experience for him to be in a school with women.

"Having had plenty of experience in both kinds of institutions, I cast an unqualified vote in favour of the co-educational. The men like to show up at their best before the women, and vice-versa. As for the outworn argument that one sex distracts the other—there is nothing in it."

In "A California Valley," however, the passages that refer to girls are glancing rather than specific. There is a long and nostalgic description of a hike by three young men to the crest of the mountains to the West: "one last strong race to the top to behold the sun go down . . . a globe of boiling gold in an opalescent sea." The hike is described as "a tramp" that the three will remember "however long their days."

Thinking that "W. F. H." and "K. A. B." of the poem's dedication might have been Bell's companions on this memorable expedition, I search out such initials in the Stanford yearbooks and directories. For each set of initials just one youth meets the alphabetical requirements: William Fraser Herron of Palo Alto, a history major who later studied law, and Karl August Bickel, an English major who became a journalist.

The friendship with Bickel began in Bell's second year when he moved from an off campus boarding house to the dormitory, Encina Hall, where Bickel, a freshman from Geneseo, Illinois, lived. In 1907 Bickel inscribed his class yearbook to Bell in memory of a time "when companionship was a red letter event in my college life."

In later years Bell enjoyed using names of friends for characters in his novels. Bickel, who ultimately became the president of the United Press, was to be the source of "Bicknell," the American journalist in *Desmond,* who advises its English hero on how to get along in America:

"Forget that you are English. . . . Eat American food and think American thoughts. Never let yourself believe that England does anything better than we do. . . . Above all choose native-born Americans for your friends. . . . For I have seen more young Englishmen go to the devil through snobbery and swellhead than through any six healthy vices combined."

(It was advice that Bell was always to follow, although I doubt that it came from the young man from Geneseo, Illinois.)

Bell's friendship with Herron developed when Bell was a graduate student and Herron a freshman. Herron, later a lawyer practicing in San Francisco, appears in *The Time Stream,* but not as an attorney.

But once again I am getting ahead of my story.

At the beginning of his last year at Stanford, like most young men approaching the end of their college career, Bell began to think about how he was going to earn a living after graduation. He took only four mathematics courses that year: Advanced Calculus (Allardice), Elementary Theory of Groups (Miller), Continuous Groups (Blichfeldt), and an "extended course" in the Theory of Functions (Allardice). But he signed up for several courses in education. One of these was designed, according to the catalogue, "for students prepared to do advanced work, the nature of the investigation being determined by the student's preparation and needs." Although the catalogue suggested "studies in administration, statistics, educational history, and the curriculum," Bell set up for himself a course in Physics for Teachers. That year he received a few letter grades: an *A* in Advanced Calculus, *B*'s in Continuous Groups, *B*'s in two of his education courses, and a *C* in Educational Theory and Practice.

At the time of his graduation, so he later wrote, he intended to specialize mathematically in Lie's continuous groups. It is to be assumed, although he never said so, that this decision was made under the influence of Blichfeldt, for such a mathematical choice was completely out of character for the mathematician that Bell was to become. Many years later, in his Prospectus for *The Development of Mathematics,* he discriminated between the two basic types of mathematical minds:

"One type . . . prefers the problems associated with continuity. Geometers, analysts, appliers of mathematics to science and technology are of this type. The complementary type, preferring discreteness, takes naturally to the theory of numbers in all its ramifications, to algebra, and to mathematical logic."

As a mature mathematician Bell belonged to the latter group. He was always to find the continuous distasteful, referring on occasion to the continuum of real numbers as "that chaotic smudge," and to be at home only in the discrete, where each element with which one deals is separate and individual, as are the whole numbers.

The summer after graduation he got a job that he later described in a letter to some young women establishing the six weeks of residence necessary then to obtain a divorce in Nevada: "The mention of Fallon (Nev) touched a sore spot. It was there back in '04 I sweated out my first job. A hellish experience—mules, whores, flies, filth and germs. Fallon was 'camp C' on the project. Within 6 weeks all the engineering staff, except me, had been eliminated by typhoid. There was no sanitation. I escaped, owing to an attack at Stanford just before I graduated. With a population of 812 Fallon was to have been a 'capital' of the large agricultural agency opened up by the canal. It died aborning, because the land was settled by white trash. The big boss of the job was (Gen.) Mead after whom Lake Mead is named."

At the end of the summer he came back to Stanford and signed up for yet another course in group theory. On his transcript it is listed as "Math 22 (Th. Groups)," but this designation must be incorrect since he had earlier taken a course with the same number and title. It may well have been some more work in continuous groups with Blichfeldt. By limiting himself to just the one course he was able to hold a full-time job at a Palo Alto telegraph and telephone company.

From the memorabilia in Watsonville there is just one other scrap of information about the two and a half years Bell spent at Stanford. In the inscription in the yearbook given him by Bickel he is identified as "Eric T. Bell (James Temple)." The choice of this nom de plume is a bit hard to explain. If James had been merely his father's name, taking it would not have been odd. But James was also the name of the older brother from whom he was apparently alienated. And while Temple was his own middle name—to do with as he liked—he must have learned from visiting his father's grave in Norwood that it was also the maiden name of his father's first wife.

A couple of hours in the library at Stanford, examining old copies of the campus literary magazine, turn up nothing by James Temple during Bell's undergraduate years. But in the fall of 1904, when he was a graduate student, there are several short poems by Temple in the magazine and one in the yearbook.

The year that Bell graduated was the year in which he reached his majority and received the inheritance coming to him from his father's estate. At the beginning of 1905, having completed the graduate course in group theory, he had "securities" in his pocket and was looking for a place to invest them. What better place than San Francisco, the fabled, magical, frisky City?

PIERRE FERMAT *"at least* Newton's equal *as a pure mathematician"*

The Forces of Nature

At the beginning of the spring semester 1905, Bell moved up to San Francisco and got a job as a teacher at The Lyceum. According to a local newspaper, this was "one of the leading accredited preparatory schools of the kind on the Coast" and had been "prominently before the public for eight years [preparing] hundreds of students for the universities, law and medical colleges, and for teachers' examinations." Headed by a Prof. L. H. Grau, the school was located in the Phelan Building, one of the triangular buildings characteristic of San Francisco, the result of a rectangular grid of streets that, ignoring hills and cliffs, intersects at Market Street, the City's great diagonal thoroughfare.

In the San Francisco Directory for 1905 Bell is listed at 729 California Street. From the number it is clear, through innocence or intention, he had established himself at the edge of the City's red-light district, for his "lodging house," according to Sanborn's insurance maps for 1905, was located on the south side of California less than a block above Dupont, the notorious thoroughfare of San Francisco's Chinatown. The house itself, however, was next to the Diocesian House of Grace Church and less than a block up California Street from Old St. Mary's Church. On the neo-gothic tower of the latter was an admonition that Bell would have passed every day:

Son, Observe the Time and Fly from Evil.

Once the cathedral of the Archdiocese of San Francisco, Old St. Mary's had lost that status as its surrounding neighborhood had become increasingly unsavory. At the time that Bell arrived, fresh up from the Stanford "Farm," Dupont Street north of California was ruled by Chinese "tongs"—for the most part criminal brotherhoods that fought with hatchets, cleavers, and knives hidden in the sleeves of their flowing kimonos and were known as "hatchet men." The street was lined with opium dens and gambling halls presided over by the "pigtailed Chinese" later featured in some of "John Taine's" novels. Although the fathers at Old St. Mary's had recently succeeded in having prostitution on Dupont south of California limited to the upper stories of buildings, the response to the new law had been to reduce the size and increase the number of the accommodations upstairs. As Bell walked down Dupont to take up his teaching duties at The Lyceum, which was at the foot of the street, where it intersected Market, he would have heard the sounds of remodeling that, so the fathers complained, disturbed the Mass even on Sunday.

Bell had come to the City "to invest" as well as to work. Possibly as a result of his earlier experience at the Palo Alto telegraph and telephone company, he joined with some friends in a partnership to start a telephone company. The company—so Bell once told Taine—laid the telephone lines for the elegant Fairmont Hotel. In 1904 there were six telephone companies listed in the San Francisco City Directory; in 1905, eight. The two new companies were the Central Energy Secret Telephone Company and the Home Telephone Company. I rather fancy the Central Energy *Secret* Telephone Company for Bell.

Some feeling of his life at the beginning of 1906 can be gained from his classic novel, *The Time Stream,* which recounts the time-travel adventures of a group of people living in the City that fateful spring. Frequent reference is made in the novel to "Sanguinetti's," the restaurant where members of the group meet for ravioli and red wine. It is listed in the city directory of 1905. Although there is not a listing for "Hochst's," the favorite saloon of the group, it too may have existed by 1906, a year for which there was to be no directory. The apartment in which the group experiences the earthquake is also recognizable, although it has been discreetly moved a block farther up California.

"You live in a good neighborhood, Culman," Herron remarked with a nod down toward Dupont Street.

"Very," Culman responded. "Two churches—Episcopal and Catholic—in one block."

"Vice and virtue rubbing shoulders," Palgrave murmured. "I often wonder which it is that first attracts the other."

"A mad city," Will Irwin wrote. "Inhabited by insane people." The "downtown" was small, and almost all of it was immediately accessible to Bell from his lodgings. Two blocks above him on California were the mansions of the "nabobs," who gave the hill its name "Nob Hill," and two blocks below, on Montgomery Street, was the financial district. The theatres and music halls were two blocks over on Bush Street and the shops, well stocked with luxuries and extravagances, began a block below Dupont, and continued to Market.

In spite of his job and the many attractions of the City, Bell continued to be active in mathematical work. But he was no longer interested in continuous groups.

"Cartan's paper, then quite recent, seemed to clean up all the things worth doing, and I turned definitely to the Theory of Numbers. This I studied entirely by myself—as usually is the case with a man's real interest."

He had chosen the field of mathematics that the master mathematician, Carl Friedrich Gauss, called the *Queen* of the subject—a field that Bell was to describe many years later, when he spoke from experience, as "one of the most difficult—although apparently the simplest—departments of mathematics." A glimpse of the apparent simplicity of the theory of numbers can be gained from one of its most famous problems, what is the wrongly but commonly called the "Last Theorem" of Pierre Fermat. (It is neither "last" nor a "theorem"—since it has never been proved.) The sense of the statement that the equation $x^n + y^n = z^n$ is not solvable in positive whole numbers when n is greater than two can be grasped by anyone who has had even a smidgen of elementary algebra. But the proof of the statement—ah, that is yet another story, which Bell himself was to recount at the end of his life under the title *The Last Problem*.

How did Bell happen to settle on number theory?

"I never asked him the question," H. F. Bohnenblust, later a close friend and colleague, responds to my query. "Probably somebody told him to stay away from it—a compelling reason. But seriously ... the abstract nature of the theory and the fact that it is far away from the rest of the world would attract Bell. ... Number theory remains an art and that would also appeal to Bell."

All one can say with certainty is that, once on his own mathematically, it was a case of love at first sight. By the beginning of 1906, at the very latest, he was studying three classic texts in number theory, the authors of which were to be his instructors in the subject: Paul

EDOUARD LUCAS
"had the right spirit"

Bibliothèque Nationale, Paris

Bachmann, who wrote in German; Edouard Lucas, who wrote in French; and G. B. Mathews, who wrote in English.

He continued to maintain contact with Stanford, keeping in touch with friends and mathematics professors, attending lectures by such visiting notables as William James, borrowing mathematics books from the university library. On February 23, 1906, he went down to Palo Alto for the meeting of the San Francisco Section of the American Mathematical Society. Although he was not a member, he was permitted to present a paper after being introduced by Professor Blichfeldt.

"Mr. Bell's paper showed that the sifting of the primes from one to infinity can be made to depend upon certain arithmetical progressions, whose first terms are the squares of the odd numbers taken in order, and whose common differences are the odd numbers, also taken in order," reported the Secretary, Bell's former professor, G. A. Miller.* "This method gives all the primes uniformly, and can be used as a starting point from which to determine the number of primes less than a given integer."

---

* Abstracts of talks were probably written then, as now, by their authors.

The paper was a variation on the ancient "sieve of Eratosthenes," the earliest method of separating out the primes, and was not of much interest, as Bell must have realized, for he never published it.

Two months later, in the early evening of April 17, 1906, he strolled through the downtown area of the City with a friend who was an engineer. The friend had been talking excitedly all that spring about the possibility of an earthquake in San Francisco, and in the course of their walk he pointed out some buildings that would be certain to fall in such a case. Returning to his lodgings, Bell worked until 2:30 a.m. "and consequently was dead to the world" when at 5:13 "the bust up came."

All this is contained in a long letter about the events of the next few days, published in the *Eagle* of the Bedford Modern School.*The story Bell used to tell Taine, however, was that the night before the great San Francisco earthquake he and some friends had been out on the town and, when "the bust up came," he thought that he was suffering from the worst hangover in the history of the world. An earthquake of 8.3 intensity on the Richter scale was shaking the City.

Tossed as if by "a terrific storm at sea," Bell did not dare to get out of bed while everything was shaking. He expected the steeple of Grace Church to fall on his house and was certain that he would be killed.

After the shaking stopped, he found that his bed had "waltzed" across the room, his bookcase had been thrown to the floor some twelve feet distant from its regular place, his dressing table was "in splinters." He got up, located his cash, dressed "as usual," and prepared to leave the house. He was most concerned about the securities in the safe of the telephone company in which he had invested. Within twenty minutes he and a friend, whom he identified simply as "K.," were on their way to the financial district.

"Magnificent buildings were twisted like corkscrews, and the sidewalks looked like crumpled up paper. In places goods stored in vaults under the pavement had been forced up 12 to 14 feet into the air. The cracks in the ground could not be exaggerated. . . . The street car tracks were twisted high into the air in all imaginable shapes."

Seven fires, ignited by fallen electric wires south of Market Street, were already raging. He and K. had to pick their way "like cats" among

---

* Inexplicably Bell's letter was identified in the *Eagle* as being by L. M. Bell, a student of an earlier day. I ask to see a copy only because I think that the two "Old Bedford Modernians" in San Francisco might have had some contact, but I immediately recognize from internal evidence that the author was, not L. M., but E. T. Bell.

fizzling wires. What wasn't burning was melting. The water mains had broken, and he would never forget the look on the faces of the San Francisco firemen—"the best in the world"—as they tried hydrant after hydrant and got only dribbles of water. The hose carts were rushed off for dynamite. In fifteen minutes, as he and K. watched, some of the tallest buildings in the City were being dynamited.

"The sight of a fifteen storey structural steel building being dynamited is very curious. It goes up rather slowly, and on the whole looks like a heap of dirt being given a vigorous kick."

The two young men made "a wild dash" for the telephone company's offices but found the new "fireproof" building already burning. They managed to get out some books, "the safe was where the Lord only knows." Suddenly there was a yell from the street, and they made out the word "dynamite."

"The way we got out of the building was nothing slow. Five minutes later it was a heap of bricks and scrap iron."

They hurried back to their lodgings. Grace Church was still standing and seemed to have suffered little damage. But the fire was rapidly approaching. They rushed to dig a common trench in the garden of the house.

K. buried his valuables. Bell was more concerned about the books he had borrowed from the Stanford library.

"... the infernal shaking began again. We didn't wait that time, and in my haste to get out of a window on to a roof twelve feet or so below, I made a boss shot, and nearly took the top of my head off on the window-frame. It was so funny, that the others stopped and began to howl with laughter, but they soon made a jump themselves when the roof began to cave in."

At noon the firefighters abandoned their fight to save the Church. Bell left K. and went off by himself to the office of The Lyceum to see if he could help "the old man." The elevator shaft of the Phelan Building was filled with bricks and timber, and the halls were knee-deep in plaster and bricks. He managed to find Professor Grau, and together they rescued some of the school records. On their way downstairs they saw bodies of watchmen and their wives, who slept in the building, being carried out.

Aftershocks continued throughout the day. There were seventeen in all—"four severe." In the evening Bell and his friends went up to the top of Nob Hill to watch the fire.

"There was not a breath of wind, and the black smoke and flames rose straight up in intense silence, except when the dynamiters got busy. Nero's little bonfire at Rome was a candle to a furnace, compared to this."

Strangely enough, none of the buildings that his engineer friend had designated as bound to fall in an earthquake had fallen. The friend though had been "smashed to smithereens."

Bell was most impressed by the way in which the people of San Francisco responded to the disaster.

"There was no running around the street, or shrieking, or anything of that sort. Any garbled accounts to the contrary are simply lies. [People] walked calmly from place to place, and watched the fire with almost indifference, and then with jokes, that were not forced either, but wholly spontaneous. In the whole of those two awful days and nights I did not see a single woman crying, and did not hear a whine or a whimper from anybody. The rich and poor alike just watched and waited, it being useless to try and save anything but a few immediate necessities, and when the intense heat made it necessary to move, they [would] get up with a laugh and say, 'Well, I guess we've got to leave the old shack now.'"

On the third day after the disaster he and his friends "bummed" their way down to San Jose to see what had happened to people there. They spent the night in St. James Park, which was across the street from the new offices of Dr. Pierce, his former guardian. On their way back to San Francisco they stopped in Palo Alto and found that many of Stanford University's unreinforced sandstone buildings had been "smashed to thunder."

"It will cost 4 1/2 millions to put it to rights, and they are starting right away."

Right away.

In the following weeks, as ruins cooled and martial law relaxed, the citizens of San Francisco began to rebuild. Unlike Grace Church, which had survived the quake but had perished in the fire, Old St. Mary's, although badly damaged, still stood. "If, as some say. God spanked the town/For being over frisky/Why did He burn the

Overleaf: DEVASTATION OF EARTHQUAKE AND FIRE on California Street some months after April 18, 1906, as seen from the upper block of California where Bell had had his lodgings (now the site of the Ritz-Carlton Hotel). Courtesy of Old St. Mary's Church.

Churches down/And save Hotalling's whiskey?" the citizens chanted. But the fire had also burned to the ground the unsavory neighborhood around Dupont Street, and it was never rebuilt.

Bell went back to the garden behind his lodging house (now totally destroyed) and dug up the volumes he had buried. The fire had been so hot that the marble side-altars of Old St. Mary's had melted, but "[although] everything else had cooked, the books were only half baked." Scorched pages can still be seen, I am told, in some books that Bell later gave to Caltech.

Like everyone else in the City, he began to put together his life. Only four of the eight telephone companies that had been listed in the 1905 directory still existed after the quake. These did not include the two "new" companies mentioned earlier.

Thousands were trying to collect on their insurance policies, which had been lost or destroyed in the catastrophe. A number of people were hired, housed, fed, and put to work by the insurance companies to help process claims. Bell may have been one of these, for he later told Taine that he had been involved in helping people collect on their insurance. He became rather disillusioned in the process. Even when policies could still be produced, the insurance companies often denied coverage on the basis (according to a statement in *The Time Stream*) "of some obscure clause that had been thought only to apply to Central America." This was a subject of much debate at the time, and not so clearcut as Bell made it sound.

Professor Grau's Lyceum moved out of the devastated downtown area and reopened. I have no reason to suppose that Bell did not continue to teach there, but he is not listed as doing so in the 1907 San Francisco directory. As soon as he was able, he began to purchase books to replace those he had lost in the earthquake and fire, always inscribing them to that effect. His copy of Mathews had survived, but he had lost his Lucas and was not immediately able to replace it. He began to purchase all Bachmann's books on number theory that he did not already own. These included *Grundlehren der neuer Zahlentheorie,* which had just been published. A quarter of a century later, reviewing a new edition issued after its author's death, he "warmly" recommended it to beginners.

"The exposition is unusually detailed, and Bachmann spares no pains to make clear his reasons for attacking various arithmetical problems in the way he does."

One of Bachmann's attractions for Bell may well have been the fact that he came straight out of the most glorious period in the history of the theory of numbers. He had gone to Göttingen the year after Gauss died and had heard the lectures of Gauss's successor, Lejeune Dirichlet, in the company of a fellow student named Dedekind. As San Francisco, like the phoenix it had chosen for its symbol, began to rise from the ashes, Bell pursued his study of the theory of numbers with Bachmann as his instructor and with the unattainable model before him of Carl Friedrich Gauss who by the time *he* was twenty-four had revolutionized number theory.

"After clearing up with my partners [in the telephone company] our obligations," he wrote in *Twentieth Century Authors*, "I quit business for good and returned to mathematics and, as a recreation, writing."

A fortnight after his twenty-fourth birthday, he attended the meeting of the San Francisco Section of the AMS and presented "A note on linear congruences," where "instead of proceeding as usual in the solution of a linear congruence in n indeterminates, [he made] use of the theory of partitions."

At the end of the summer of 1907 Bell left San Francisco for Seattle and the University of Washington, taking with him an intimate acquaintance with cataclysm and a healthy respect for what, as he later wrote, "some people who have never experienced them refer to as 'the forces of nature.'"

C. F. GAUSS: "Archimedes, Newton, and Gauss . . ., it is not for ordinary mortals to attempt to range them in order of merit."

Fateful If Not Fated

In September 1907 Bell found himself on a magnificently forested campus with a few modest buildings and a panoramic view of the most spectacular range of snow-covered volcanic peaks in the United States, some extinct but many only dormant.*

The University of Washington was concerned primarily with filling the vocational needs of the children of the voters of the region, but it had dreams of becoming a true university, a gathering of scholars. The dream and the reality were exemplified in the Mathematics–Astronomy Department, where Bell was now one of two graduate assistants. Its German-born head, R. E. Moritz, was a forward-looking scholar who tried to instill in his teachers an interest in pure research despite the heavy elementary teaching load they had to carry to meet the mathematical requirements of students in Engineering, Pharmacy, Forestry, Fisheries, ... The second Tuesday evening of each month was regularly devoted to a meeting of the Mathematics Journal and Research Club, the purpose of which was to review current literature and to discuss the research of the members. These included advanced students as well as faculty. The mathematical library was good, and getting better.

Even at Washington, in spite of his continuing interest in mathematics, Bell appears to have had no immediate ambition beyond quali-

---

* From 1914 to 1917, when Bell was back again at Washington, the Northern California anchor of the Cascade chain (Mt. Lassen) was producing a series of violent explosions and glowing avalanches—another close at hand instance of "the forces of nature" that were to be a hallmark of his later science fiction.

fying for a job as a mathematics teacher in a public high school. He took only two courses, which although they were listed under mathematics were in fact education courses. He may have chosen them because the mathematics courses offered at Washington were generally on a more elementary level than those at Stanford. The only ones that might have interested him, being at least on new topics, were not offered the year he was there. Still, it seems odd that having already taken a course in the History of Mathematics at Stanford he would sign up for yet another at Washington ("Teachers Course I"). The second semester he signed up for Teaching of Mathematics ("Teachers Course II"). These were the only courses he took during the year. It thus comes as a surprise to find him listed in the Washington catalogue of 1907–1908 as the instructor for an upper division course in the Theory of Numbers: "A presentation of the matter in the first five and seventh sections of the *Disquisitiones Arithmeticæ* of Gauss, as modified by later writers."

The material for Bell's lectures can be assumed to have come from his intensive study of Bachmann's book on the elements of number theory, which he had purchased earlier that year in San Francisco. This was the opening volume—although not the first to be published—of the comprehensive presentation of the theory of numbers to which its author was devoting himself in his chosen retirement. The publication of the *Disquisitiones* in 1801, when Gauss was twenty-four years old (Bell's own age in 1907), had given an entirely new direction to the theory of numbers. What had been in the time of Fermat and Euler a miscellaneous collection of unconnected, although very interesting, results had immediately become, thanks to young Gauss, a unified, coherent mathematical subject on a level with the other three great areas of mathematics: geometry, algebra, and analysis.

The subject of Bachmann's opening volume on the contemporary state of number theory had been, quite logically, an exposition of the contents of the first five books of the *Disquisitiones*. In a treatise on cyclotomy and its applications to number theory, Bachmann had earlier treated the seventh and last section of the *Disquisitiones*, in which Gauss had presented his dazzling work on cyclotomy, or the division of the circle into equal parts with straightedge and compass alone. (The sixth section, which Bell omitted from his lectures, applied the theory of the first five sections to a number of special cases.) It was Bell's later, considered opinion that Gauss's "flawless work of art" was "accessible"—although not in the concise form in which Gauss had presented it—"to any student who has had the usual algebra."

In addition to his lectures on number theory, Bell also "helped out" in the Mathematics-Astronomy Department, teaching a section of the course in Plane Trigonometry, required of all freshmen in the College of Liberal Arts, and another section in Plane Geometry and Higher Algebra, designed primarily for engineering students. He is also listed in the catalogue among instructors for Analytic Geometry and Higher Algebra, Solid Geometry, and Analytic Geometry for Engineers.*

Although the University of Washington was essentially a "street car college" with most of its students coming from Seattle and its immediate environs, there was an active college life of athletics, fraternities and sororities, clubs for almost every interest, a newspaper, and a yearbook. The last named provides the only other references to Bell's activities during 1907–1908. He is listed there as a member of the Executive Committee of the newly formed Philosophical Club, the purpose of which was "to conduct investigations of broad philosophical subjects," and was apparently one of its organizers. The other reference is in the yearbook's "Stings":

> The wily E. T. Bell
> Teaches mathematics.
> Just because in Calculus
> And in Analytics,
> Circles sometimes osculate
>
> There is no reason that, when late
> With his "lovie"
> On the campus tête-à-tête
> He should circles imitate.

On June 17, 1908, Bell and thirty other students were awarded the master's degree. Officials now at the University, examining Bell's transcript, are at a loss to explain the action of their predecessors in awarding such a degree to him. They cannot imagine from the data before them on what basis or in what field he could have earned it. In the catalogue for 1907–1908 the requirements for the master's degree are explicitly set out:

1. At least one year's work must be done in residence in undivided pursuit of the studies elected; or not less than two years

---

* Washington was a coeducational institution, and the courses in the Mathematics-Astronomy Department were specified as to the sex of students permitted to enroll: men, women, and men and women.

in residence, if the candidate is employed as a teacher or regularly engaged in any other occupation or profession. . . .

2. The candidate must elect a major subject and either one or two minors. He must earn not less than 32 credits, at least one-half being in the major subject, a part of which shall consist of a thesis embodying independent, though not necessarily original research . . . .

Bell did at least meet the requirement of a thesis, although I learn this only later in a footnote to a paper, "The equation $ds^2 = dx^2 + dy^2 + dz^2$," which was not published until 1920. There he explained that the substance of the paper was an unpublished A.M. thesis presented to the University of Washington. In his treatment of the equation of the title, which plays a prominent role in differential geometry, there was already a hint of what would be a characteristic approach for him, the invention of a general device that permits a number of previous disconnected results to be obtained.

There is no question that by the time Bell received his master's degree, whatever the subject for which it was awarded, he was an experienced teacher of mathematics. But in spite of his qualifications and the efforts that the University of Washington made to place the teachers it trained, he did not obtain a position for the school year that began in September 1908.

At this point there is another hole in the chronological net with which I am trying to catch the young E. T. Bell. In his 1931 letter to the Bedford Modern School's *Eagle* he wrote that after leaving the University of Washington he "returned to San Francisco to make enough for the last lap," but there is no documentary evidence that he was in San Francisco after 1907. In *Twentieth Century Authors* he described himself as working at different times as ranch hand, mule skinner, and surveyor but he gave no dates. He also sometimes told people that when he lost the first joint of the thumb on his right hand he was working in a lumber mill. In his unpublished novel *Desmond* there is a great deal about lumbering in the American northwest, although nothing about actually working in a mill.

Finally I find him surfacing in Northern California in the little mountain town of Yreka, the county seat of Siskiyou County. Earlier the town had been an active center of goldmining, but those days were long gone, and the chief activities were currently lumbering and farming. Under "Local Happenings" on January 6, 1909, the *Yreka Journal*, which boasted that it covered the county "like the dew," carried an item that "E. T. Bell of San Jose has succeeded H. H. Owens in the Siskiyou County High School." It added that Mr. Bell was a graduate of Cambridge University of England and has taken a postgraduate course at Stanford University. (Later, in the official Siskiyou County High School Yearbook, Bell's qualifications were accurately listed: "A.B. Stanford, A.M. University of Washington.")

The next mention of Bell in the *Journal* was a report on September 2, 1909, that he was back in town "to arrange his work for the coming year." By this time, after only one semester, he had clearly established himself among the high school teachers. At the annual Teachers' Institute (four days of instruction and entertainment for the county teachers at the beginning of the new school year), he gave one of the three talks presented to the high school section. Taking as his subject "Algebra and Arithmetic," he discussed "the preparation of eighth grade students, how their transition into high school could be smoothed, and methods of presenting those aspects of the subject that were common to both levels." The next day he spoke on "Some Suggestions Regarding School Courses." He talked "of wasted time, of the age of graduation and years spent in school, of the ordinary course being for the 'average' child [and] not for the specially bright one, of the bad combinations made in some high schools."

Siskiyou County High School, the larger of the two high schools in the county, began that school year with a faculty of seven and a student body of seventy-nine. (Many more students would attend the high school, according to the local newspaper, if there were living accommodations for them in the town.) One of the seven teachers was Mrs. Jessie L. Brown. She was a widow who had been teaching art and commercial subjects for the past two years in Shasta County, and she had been hired as a replacement for a teacher who had resigned just before the semester began. The tiny, copper-haired Mrs. Brown took lodgings a short distance from the high school at a boarding house on Oregon Street, where Bell was already established.

At this fateful moment Bell was twenty-six and Jessie Lillian Brown a few months short of thirty-three. In later years she rarely spoke about her life before she met him, and Taine learned almost as little about his mother's early life as he learned about his father's. What he did know of her family background had come purely by chance. While his daughter Laurie was a student at the University of Colorado in Boulder, she happened to see the name "Marinus Smith" on a commemorative plaque at the University and made the connection with

MARINUS "University" SMITH
grandfather of Jessie Smith

her great-great-grandfather because of the unusual first name (which
she had seen in the brief genealogy Bell had dictated to Janet before
his death). Later, visiting Laurie at school, Taine and Janet researched
the life of Marinus Smith in dusty volumes in the basement of the university library, and learned the following:

Marinus Smith had been known during his lifetime as "University
Smith" because he had given so much land to the University of Colorado. Taine's mother, Jessie Lillian Smith, was the daughter of Walter
Hubble Smith, the oldest son. At some point Walter took over the family's express business, leaving Marinus to tend his orchard lands, and
moved to southwestern Idaho with his wife, Suzie Connor, a former
schoolteacher. They had three children, Jessie Lillian being their only
daughter. She married a man named Brown and, later, E. T. Bell. This
was all the information Taine had about his mother except for a drawing among Bell's memorabilia with a notation on the back that it had
been "Done by Toby when she was in San Francisco in 1903." Taine had
never learned the full name of his mother's first husband or the nature
of his work or the cause of his death.

Because Mrs. Jessie Brown, later Bell, was to play a more important
role in Bell's life than any other person, I feel it is important to find out
who she was before she met him. For a while it seems that learning

about someone whose maiden name was Smith and whose married name was Brown is going to be even more difficult than learning about Bell himself. Then a transcript from the University of Washington, which she attended while Bell was on the faculty there, provides the information that she was graduated from Payette (Idaho) High School in 1896.

Payette—the inhabitants place the accent on the second syllable— lies in the lower valley of the Payette River in the southwest corner of Idaho. The foothills of the Squaw Ridge Mountains are to the northeast and the Snake River, which separates Idaho from Oregon, to the west. On a vacation trip through the northwestern United States in 1991 my husband and I take the opportunity to detour through Payette.

Today, according to a sign at the entrance to town, Payette has 5,448 inhabitants, about five times the number it had when Jessie Smith attended high school there, but it is still a small town with lots of stores where one can buy furniture and farm implements but very few places where one can buy dinner. Beyond the main thoroughfare there are gracious old homes surrounded by lawns and tall old trees.

The town came into being in the early 1880s as Boomerang, a construction camp and storehouse for the Union Pacific. Later, having

JESSIE LILLIAN SMITH, Class of '96, Payette (Idaho) High School

grown more permanent, it took the name of Francis Payette, a French trapper who rebuilt Fort Boise as a trading post for the Hudson's Bay Company. Jessie's father may have come to town with the opening of the railroad in 1884, for in 1885 he was reported in the local paper as having purchased a hundred acres of land in the county at three dollars an acre.

By 1891, according to a local history, there were two general stores in Payette and another ready to open, two drug stores, two livery stables, two hotels, one hardware and stove store, and a bank. There was no library but there was a new school, the building of which had been financed entirely by private donations, a matter of great pride to the inhabitants. The school was a two-story brick building of four rooms, the two rooms on the upper floor being devoted to grades seven through twelve. A "professor," obtained from the East, had introduced "the graded school" and "modern teaching methods." Jessie Smith's class was the second to be graduated.

Two years after her graduation from high school she was reported in the *Payette Independent* as having returned from Portland, Oregon, "where she has been taking a thoroughly commercial course in the Portland Business College." Her name did not appear in the *Independent* again until May 10, 1900, when there was a note that Miss Jessie Smith, who was in San Diego, California, with her mother, had been informed that she would have to go to Colorado to look after some property she had inherited from her grandfather.

At this point there is a discontinuity. Sometime between May 1900 and 1903 Jessie Smith became Jessie Brown, and Mr. Brown died. In the San Francisco city directory for 1903, "Jessie Brown, Widow" is listed at 758 Golden Gate Avenue. Even though her name is a very common one, she is identifiable there on the basis of Bell's notation that she was in San Francisco in 1903.

In addition, on the roster of the Yreka faculty she listed herself as having been trained at "Best's Art School." There is no Best's Art School in the directory in which her name appears, but under "Portrait Artists" there are listings for Mrs. Alice M. Best, Arthur Best, and Harry C. Best—all at the same address on Market Street. But who was Mr. Brown, and how did he happen to die so soon after the marriage?

The names of five counties come up in the course of my research into the background of Jessie Brown, but in not one of these counties is there a record of a marriage between a Jessie Smith and a man named Brown. Among the books Bell left in Watsonville there is a copy of

Les Misérables with the flyleaf torn out. On the corner of the inside cover, the name "J. O. Brown" has been written in pencil, possibly even by Bell. In written records and in conversation he always made the point, often when it seemed unnecessary, that his wife was a widow when he married her. For this reason Janet had the impression that her mother-in-law was a divorcée. On the certificate of her marriage to Bell, however, she is listed as "Jessie Lillian Brown, Widow."

Some years after his marriage, Bell wrote a novel, never published, titled *Red and Yellow.*

"Married couples," it began, "sometimes like to look back over their lives and compare notes. ... Then they fall to wondering what fate, what chance, ever wove the distant threads of their lives into a single strand. How in the name of reason was it possible that they should ever meet? Was it ordained from the beginning of time? They almost persuade themselves that it was."

A teacher resigned just before the term began, a teacher from a neighboring county was hired as a replacement. Thus occurred the personally fateful, if not fated, meeting of E. T. Bell and Jessie Lillian Brown during Yreka's "institute week" for teachers in September 1909.

JESSIE LILLIAN SMITH at twenty (1897 or 1898)

128

The Decisive Event

The old Siskiyou County High School where Bell and Mrs. Jessie Brown taught burned to the ground in 1916, and the bulk of its records were destroyed; however, Eleanor Brown and Anita Butler of the Siskiyou County Museum are able to find information for me about their teaching in contemporary yearbooks.

In 1909 the high school offered a variety of courses: the Classical (which included Greek), the Literary, the Scientific, and the Commercial as well as a combination course of Natural Science and Commerce. Each of these prepared its students specifically for certain colleges at the state university—the Scientific Course preparing "for the Colleges of Agriculture, Mechanics, Mining, Civil Engineering, as well as for the Colleges of Pharmacy and Medicine." The principal of the high school, who had an A.B. from Harvard, taught Physics, Solid Geometry, Trigonometry, Geometrical Drawing, and Greek. To Bell were allotted classes in Algebra and Plane Geometry along with Botany, Chemistry, and Agriculture.*Mrs. Brown taught "the Commercial Branches"

---

* Bell kept abreast of developments in this seemingly alien subject. In *Science* (September 30, 1910) he published a short review in which he summarized a paper, "The Influence of Nutrition upon the Animal Form," presented at the meeting of the Society for the Promotion of Agricultural Science, "[which] may not otherwise come to the attention of the experimental morphologist and others to whom it may be of considerable interest." After a commendably succinct summary he concluded with a paragraph that hints of his future interest in *devolution*: "It is interesting to note that the ancestral type, from which the modern beef animal has been derived, corresponds in the shape of the skeleton to the underfed animals described above. . . . The narrowing of the skeleton in response to an inadequate food supply may be a physiological adaptation, or it may be a case of reversion."

and "Free Hand Drawing." The following year "Domestic Science (Cooking)" was added to her schedule while Bell was promoted to Advanced Mathematics and Physics in addition to the three sciences he had taught the previous year.

There is also at the Siskiyou County Museum some description of Bell's effectiveness as a high school teacher. *Houses That Talk* by Fred and Bernice Mecomber (1986) contains interviews with descendants of pioneer families. One is with Ethel Ackerman (since deceased), the youngest daughter of the owner of Ackerman's Drug Store.

"Miss Ethel graduated . . . in 1910 as president and valedictorian of her class and upon recommendation of her mathematics teacher, Eric Temple Bell, enrolled at the University of Washington in Seattle. He had taught there himself, and he knew the professor who would be Ethel's teacher. She signed up for engineering-mathematics, and the professor [at Washington] . . . told her to go ahead and enroll, but she would promptly 'flunk out the first semester.' But Ethel was determined after that to show him . . . , and [one assumes, with the benefit of Bell's instruction] she did."

The yearbook for 1909–1910 contains two unsigned sketches of "Mr. Bell" in the section labeled "Joshes." Although not specifically identified, they are placed next to exchanges that involve him and are recognizable by the aquiline profile. They show that he was still wearing a beard and that he was a relaxed teacher.

There is unfortunately no report in the archives of the Siskiyou County Museum of the developing relationship between Bell and the tiny copper-haired widow who taught art and commercial subjects, who smoked, spoke her mind, and dressed in what other women described as a "bohemian" manner. The spring following their meeting, the U.S. census taker found them living with two other boarders in the home of Mrs. Olive E. Auteurieth at 806 Oregon Street. Mrs. Brown's responses to the census taker's questions about her family background are consistent with what I know about her, but Bell's are a different matter. He gave Scotland as his birthplace, but to questions about his parents he replied that his mother was born in France and his father in Malaysia. The census taker wrote down "Malaya," then crossed that out and wrote "Asia."

Something of the courtship of the two can be glimpsed from stories they later related to friends. The romance apparently developed in the manner favored by writers of romantic comedies: constant attention-getting sparring between hero and heroine until they fall at

Mr. B. (describing how the surface of a liquid would look if
magnified until the atoms were visible):

> ''And the rocket's red glare,
> Bombs bursting in air,
> Gives proof through the night
> That the atoms are there.''

last into each other's arms. She made outrageous remarks, calling to
him when he crossed the street to avoid her after a spat, "What's the
matter—are you afraid you are going to lose your virginity?" He was
equally outrageous, demanding when she appeared at breakfast with
a rash on her chin, "Is it syphilis?" But the wistful expression on her
small face when she was dejected reminded him of the look on the
face of the little dog that Dickie Doyle had immortalized on the covers of *Punch,* and he began to call her "Dog Toby." She responded by
calling him "Dog Romps."

In spite of their rowdy exchanges, they took each other seriously.
She was to be the model for his heroines—independent, adventurous
girls who were not afraid to be unconventional when necessary—girls
who never hesitated to tell their swains what to do and were always
the first to recognize when the hero was falling in love with them—and
told him so.

During the year 1909–1910, the first in which Bell and Mrs. Brown
taught together, there was a civic project that was of interest to both of

them. Andrew Carnegie, the Scottish-born steel magnate and philan-
thropist, had offered to match local funds raised for libraries on the
condition that the libraries would bear his name. At the beginning
of the school year prospects for a Carnegie Library in Yreka were not
good. A reading room recently established "in the comfortable and
well located offices of the Chamber of Commerce" had been closed by
the editor of the *Journal* (the chairman of the library committee) because, as he wrote, "the dear people" of the town did not use it.

"The time will come when a library will be patronized here," he
predicted, "but that time is a long way off."

In spite of this pessimistic assessment, a group of citizens that included Mr. Bell and Mrs. Brown took up the campaign; and a few years
later, a Carnegie Library did open in Yreka. Bell always spoke with
special pride of their role in the campaign for the library.

I can give scarcely any specific information about what Bell was
doing in mathematics at this time. I can say that he was continuing to
pursue his study of that subject. In 1909, in Yreka, he had finally been
able to replace his copy of Lucas's *Théorie des Nombres*. More than for
Bachmann or Mathews, he had a personal as well as a mathematical
affinity for Lucas.

"Having mentioned Lucas," he was to write in *Mathematics, Queen
and Servant of Science*, "I may recall that he was another of the great
amateurs, in the sense that, although he was conversant with much of
the higher mathematics of his day, he refrained from working in the
fashionable things of his time in order to give his instinct for arithmetic
free play. His *Théorie des Nombres* . . . is a fascinating book for amateurs
and the less academic professionals in the theory of numbers."

But in spite of Bell's continuing study of mathematics and his early
presentations to the San Francisco Section of the American Mathematical Society, he had published not a single mathematical paper.

He continued to write poetry, also without publication. (One of
the nonmathematical books he bought in Yreka as "A gift from his no-
ble self" was the collected works of Shelley.) In the spring of 1910, carrying a heavy load of at least five classes a day, treating five different
subjects, as well as participating in extracurricular and civic activities,
he began to write an epic poem that he described as "about time." *The
Scarlet Night* ran eventually to more than seven hundred typewritten
pages. Revising it forty years later, he told the poet Muriel Rukeyser
that he wrote the entire poem in the spring and summer of 1910.

He described the story as originating in two dreams.

"The first was the result, no doubt, of an attempt to answer this question: 'what would be the most disturbing thing a sane mind could experience, say just before crossing the line between sanity and insanity?' In the dream the answer was given, symbolically, as an existent contradiction of the accepted laws of reason, specifically the 'law' which asserts that nothing can both be and not be . . . ."

The second dream was later treated again in a shorter poem about the violent destruction of "the temple." Titled "The Temple Steps," it appeared in *Recreations*, a collection of poems privately published by Bell in 1915, with a footnote explaining that it was an attempt "to describe a dream of extraordinary vividness which the writer had had four times at intervals separated by from one to ten years."

The Scarlet Night has a dreamlike quality, the action deriving from myths and legends that seemed to Bell to have been at least partly responsible for some of the religions of the past. In a letter to Mrs. Rukeyser he described these as "fossils of thought," some of which are "so deeply embedded in our subconscious minds that we are unaware of their existence till we try to analyze the symbols in terms of which abandoned beliefs about the nature of the universe were preserved . . . ."

While this explanation of Bell's as to the origins of *The Scarlet Night* was written forty years later, I am placing it here in the narrative of his life because it throws considerable light on ideas that were ultimately to run through his science fiction novels and even some of his popular books on science.

"The main root of the story . . . ," he continued to Mrs. Rukeyser, "is in that fossil of thought concerned with the legends of time and attempts to answer the question 'what is time?' or to show that the question is improperly put. It need hardly be said that no answer to this question is attempted here. But one thought-fossil which may be the remains of [an] attempt to make sense of the question has survived for thousands of years in the idea of the Eternal Recurrence which continues to attract the speculative mind."

Ouroboros, the snake devouring itself, is the dominating symbol in the poem, as it was later to be in Bell's novel, *The Time Stream*.

"In the older mythology [this symbol] has been interpreted as a representation of a cyclic transformation of matter into energy and vice-versa, also, more suggestively, as the symbol of a closed, circular time with its eternal recurrence," Bell continued to Mrs. Rukeyser.

"How, if there were such a time, could the serpent be cut, and contin-
ual change, not returning on itself, be made possible?"

This is the problem that concerns the three main characters of *The
Scarlet Night*.

"The action is on three interconnected levels: the now, as pictured
in the Babylon of the Captivity . . . ; the second level is the past, as
remembered in dreams and visions by the three principal actors; the
third level is the future after the serpent is cut—if it is."

Shortly after Bell completed *The Scarlet Night*, he came across a little book, published just the year before, in which he found articulated
many of the ideas implicit in his poem. The book was *The Interpretation
of Radium* by Frederick Soddy, who later won the Nobel Prize for his
theory of isotopes but was at that time a lecturer in physical chemistry
and radioactivity at the University of Glasgow. Bell was fascinated especially by the last chapter of *The Interpretation of Radium*, a chapter
that Soddy decided to omit from the second edition of the book.

"How strangely then some of the old myths and legends about
matter and man appear in the light of our recent knowledge [of ra-
dium]," Soddy had written.

"Can we not read into [the traditions and superstitions that have
been handed down to us from prehistoric time] some justification for
the belief that some forgotten race of men attained not only to the
knowledge we have so recently won, but also to the power that is not
yet ours? [But] one can see also that such dominance may well have
been short-lived. By a single mistake, the relative position of Nature
and Man as servant and master would, as now, become reversed, but
with infinitely more disastrous consequences, so that even the whole
world might be plunged back again under the undisputed sway of Nature, to begin once more its upward toilsome journey through the ages.
The legend of the Fall of Man possibly may indeed be the story of such
a past calamity."

But I am getting ahead of my story again.

Back in Yreka in the summer of 1910, in spite of the time that he
was spending on *The Scarlet Night*, Bell did not abandon mathematics.
The *Journal* reported on July 27, 1910, that E. T. Bell of the High School
was in Palo Alto and on August 24, 1910, that he was in Berkeley, "using the University Library for mathematical work." During the years
in Yreka he had become familiar with the writings of C. J. Keyser, the
chairman of the mathematics department at Columbia University, and
had found them attractive, not just because of Keyser's philosophical

CASSIUS JACKSON KEYSER
Columbia University

approach to mathematics, but also because of his literary style. Keyser was a mathematician who could write of "sensuous" curves. His papers showed, in Bell's view, "that mathematics was more than an interminable grind of theorems." When he heard that the following summer Keyser was to lecture at Berkeley, he asked a friend in that area to attend and to take notes for him.

Mrs. Brown was reported as spending the summer in San Diego with her mother, and Bell went down to visit her there and to meet her mother. Among his paintings in Watsonville there is a small landscape labeled "Ocean Beach (San Diego) August 1910 (The first watercolor I ever did. Toby directed. We sat looking up the valley.)" It is a drab thing, completely different from his later bold and vivid works in the same medium.

Although the newspaper report of the Teachers Institute in September 1910 is not so detailed as in 1909, it is clear that in the year and a half just past Bell had become the leader of the teachers in the county. It was he who was chosen to read the lengthy Resolution of Thanks from "First, That the present Institute has been exceptionally enjoyable, helpful and instructive . . ." to "Ninth, That we the teachers of Siskiyou County will earnestly endeavor for the period of one year . . . to be human beings."

Bell himself was very busy being a human being, and colleagues and students could not have failed to note the developing intimacy between him and Mrs. Brown. From the yearbook report of 1910–1911, however, it does not appear to have interfered with their teaching.

"This year pupils desiring to do so have studied advanced chemistry and advanced mathematics under the able direction of Mr. Bell. The amount of work accomplished has astonished the office. It is believed that the young people taking these courses have not only prepared themselves to take advanced work in college and university, but to utilize their knowledge in practical ways should they carry on their schooling in no higher educational institution."

The teaching of drawing also came in for praise, although without mention of Mrs. Brown.

"The older pupils of the school have assured the office that the work in the drawing room this year is far in advance of anything heretofore done in the art line in this school."

Mr. Bell and Mrs. Brown took long walks together, their destination being a small hill outside town from which there was a view of Mt. Shasta, which was not visible from the town itself. The mountain, one of the dormant volcanoes of the Cascades, is a magnificent one, 14,162 feet above sea level, its majestic snow-covered cone rising abruptly some ten thousand feet from the surrounding countryside. I would like to be able to write that there, sheltered by a stand of rocks beside which a sapling struggled to grow, Bell read aloud from *The Scarlet Night*, and Toby conceived the ambitious project to which she was to devote many years—a book of colorful and illuminated woodblocks within the borders of which his poem would be copied out in red ink. But I can write only that it was here, at this spot, that Bell "popped the question," as he put it, and she said, "Yes."

The marriage took place the day before Christmas 1910 in the home of the English and music teacher, Miss Josephine Colby. The officiating clergyman was the Rev. J. M. Wright, the rector of St. Mark's Episcopal Church. According to an account in the *News*, Miss Colby's home was "tastefully decorated" with Oregon grapes and Japanese maple, the bride carried a "beautiful" bouquet of violets, and the ceremony was followed by a "tempting" wedding breakfast.

It should be noted that in marrying Bell, who was five and a half years her junior, Toby was flouting a well-established convention of the time that a wife should be younger than her husband, usually quite a bit younger. In addition, according to a recent statute, she was forfeiting her American citizenship by marrying an alien—even though she continued to reside in the United States.*

No member of either family was present at the marriage, and Bell always insisted that he never informed his family that he had married. Yet his marriage to "Toby" was the decisive event of his life, and of his mathematical career. He had been drinking heavily before he met her and was "on the skids," as he always said, usually adding that she had picked him up "out of the gutter."

The next September, when the Siskiyou County Teachers' Institute convened, Mr. and Mrs. E. T. Bell were not present. They were on their way across Canada by rail, retracing the journey Bell had made when he returned to America in 1902. From Montréal, they planned to go to New York City. There they intended that Bell, on the basis of mathematical work he had been doing, would get a Ph.D. from Columbia University in the coming year.

---

* The statute (1907) was modified in 1922 but continued in force for marriages to certain specified aliens. For this information I am grateful to Linda Kerber's article "The Paradox of Women's Citizenship," *American Historical Review,* April 1992.

E. T. BELL, twenty-eight, as a graduate student at Columbia

A Lusty Old Subject

When E. T. Bell, twenty-eight years old, presented himself at the office of Cassius J. Keyser, the chairman of the Mathematics Department at Columbia, he had behind him remarkably little formal education at the university level. There had been none at all in England—only two undergraduate years and one graduate course at Stanford—and then that rather odd master's degree at Washington based on "Teachers Course I and II" and an unpublished thesis. On his application for the Degree of Doctor of Philosophy, he embellished this academic record a bit by listing under "Previous academic and professional training," his printing so small that it is very difficult to read, "Univ. of London 1900–01."

There is an account of the meeting in a memoir Bell later wrote for *Scripta Mathematica*. From a sketch of Keyser lecturing at Berkeley, which he had received from his friend, Bell recognized the professor the moment he stepped out of his office. He introduced himself, and the professor inquired as to what he expected from Columbia.

Bell stated the blunt fact, a Ph.D.—any education that might accompany the degree would be incidental.

The professor observed that a good many people would like to get a Ph.D. degree from Columbia, adding with emphasis, "*and a good many don't get one.*" How long, he inquired, did Bell expect to spend in acquiring a Ph.D.?

"One academic year."

"And you are confident you can do it?"

"The only thing I am confident of," Bell replied, "is that my money won't last a day over ten months."

Although Keyser remained doubtful, he gave Bell some advice on courses he should take and approved his candidacy.

Bell took two courses from Keyser himself, one on Modern Theories of Geometry and the other on Whitehead and Russell's *Principia Mathematica*, the first of three volumes that were planned to translate all mathematics into a new symbolic language and thus eliminate logical difficulties that had begun to appear in the foundations of the subject. The professor thought that Bell's mind would be broadened by contact with such minds as those of Whitehead and Russell. Years later, after Bell had heard many of the best scientific lecturers in the world, he was to say, "I never heard a better than Keyser in his prime. Exposition was with him an art. In his union of severe mathematical rigor and disciplined poetic imagination he was unique."

In addition to his courses with Keyser, Bell signed up for the Theory of Functions of a Complex Variable (Fiske), Theory of Invariants (Cole), and Elliptic Functions (Maclay). Although the last named was his first formal course in that subject in which Mr. Langley had provoked his interest so long ago at the Bedford Modern School, he had already collected a number of books on function theory and had purchased two volumes specifically devoted to elliptic functions.

As a candidate for a Ph.D. degree he also had to choose two minors. The most frequent combination was a topic in physics and a subject in some division of mathematics not too closely connected with the subject of the student's dissertation. Bell chose again to take the History of Mathematics, this time from Professor D. E. Smith, the doyen of American historians of mathematics. Since Smith's course was Bell's third in the subject, he was becoming an authority on its teaching.

"[At that time] some of the dreariest ramblings ever endured in university lecture rooms by bored students earning a fairly easy credit, were those perpetrated ... by professed historians of mathematics," he was later to write in a volume dedicated to Smith. "These well-meaning and unimaginative men transferred to history the pseudo-scientific fatuity of accuracy to the sixth decimal long after a rapid succession of basic new discoveries had outmoded profitless meticulosity in science."

Professor Smith was not that kind of historian. His course was one of the few "that did not extinguish interest" in the subject. In addition—as a result of Smith's presence—there were at Columbia several remarkable mathematical collections of instruments, manuscripts, and early editions of classic works. Among these was a copy of the

Arithmetica of Diophantus that had reproduced in its margins the notes that Fermat had made in his own copy of the *Arithmetica*. Bell felt privileged to handle this book, and he spent as many hours as he could in the library poring over Fermat's notes.

(Late in life, when Bell described in *The Last Problem* the effect that contact with the *Arithmetica* had had on the mathematical life of Fermat, he resorted to the metaphor of an earthquake, the slipping of the opposite sides of a fault in the earth's crust: "After the slip the countryside may be quite different topographically from what it was before. . . . Fermat's mathematical life was like that. The fault which rearranged it occurred in his early thirties [when a copy of the *Arithmetica* fell into his hands]. Thereafter . . . he moved at once into the first rank of creative mathematicians, where he has remained ever since.")

D. E. SMITH, Teachers College, Columbia University

For his second minor Bell selected two English courses—the History of the English Language and English Composition. Admission to the latter was restricted, and students applying were required to submit some original composition "as evidence of proficiency." He probably submitted a selection of his poems, or a portion of *The Scarlet Night*. There is no indication that he had yet tried his hand at writing fiction or popular scientific exposition.

In spite of Keyser's doubts about Bell's ability to obtain a Ph.D. within the year, the young man moved ahead confidently, having (it seems) completed most of the work before he arrived. In November he submitted his topic: "The Cyclotomic Quinary Quintic." The topic was approved, and in April the completed dissertation was accepted.

At this point in my search for E. T. Bell I find myself, for the first time, up against the mathematician Bell. The dissertation is a formidable and specialized document firmly planted in nineteenth century mathematics because that was the mathematics that he had studied on his own. It is long—almost a hundred printed pages—and page after page is filled with complicated computation. What does it all mean? I question some number theorists to see if they can tell me. The theory of numbers is a relatively small field of mathematics, but nevertheless highly specialized. The dissertation keeps coming back: "This is not really my field," or even "This is not my quintic." John Brillhart of the University of Arizona writes most frankly, "It does seem to touch ground, but I would have to study it some to see what he does. It is written in that wretched style [of the nineteenth century] in which the word 'theorem' is never used, and one can't distinguish the silk purses from the sows' ears. Anyway, what may be there is so well disguised by bad writing that to *really* read it would be taxing at the very least."

I turn for help to Albers and Alexanderson. They have some responsibility, I point out, for involving me in a project that has turned out to be quite a bit more than I anticipated when I agreed to take it on. Even if, as I still plan, I limit myself to writing only about the young Bell, I am going to have to treat a dozen years of mathematics. What am I to do?

Albers recalls that while researching Bell's life in 1981 he interviewed Lincoln Durst, at that time Deputy Executive Director of the American Mathematical Society. Durst had taken his doctorate under Morgan Ward, Bell's first doctoral student at Caltech, and had also known and been very fond of Bell himself. In fact, after Bell's death he had voluntarily assumed the quite considerable task of compiling a Bell bibliography. (With mathematical papers, general articles, and reviews in addition to the many books, it had run to more than three hundred items.) Albers suggests that Durst, who has since retired, might be willing to take a look at the dissertation. Alexanderson, who is one of the mathematicians I have already consulted, promptly volunteers to send it to Durst the next day.

If I had conducted a person by person search of the mathematical community, I could not have found a mathematician better suited to help me than Lincoln Durst. In addition to his mathematical connections to Bell, he has a deep affection for "Romps," as he still refers to him, and a desire to see that he receives credit for results that are still being "discovered" by other mathematicians who, as he says, "should know better." Not only does he sit down and study Bell's dissertation and the mathematical background out of which it came, but he also goes on to organize and evaluate the more than two hundred mathematical papers Bell published. What I say about Bell's mathematical work in this book, unless it is attributed to some other mathematician, is based on Durst's many subsequent communications.

Bell's dissertation is, Durst confesses, a tough assignment. Part of the difficulty is that what Bell was trying to do would now be done in a much different way, and in a much different language. In his dissertation he was going back to an already venerable subject: cyclotomy, or the division of the circle into equal parts. The subject had developed out of Gauss's solution to the question of which regular polygons can be constructed by straightedge and compass alone—a problem originally posed by the Greeks. Although it would be "grossly misleading," as Durst points out, to suggest that cyclotomy is concerned primarily with cutting up circles and constructing polygons, I shall treat the subject in those terms for the purposes of this narrative.

Bell's interest in cyclotomy went back at least to 1907 when, in the year following the San Francisco earthquake, he added to his Bachmann collection the volume on cyclotomy and its applications to number theory. In that work Bachmann naturally devoted a great deal of space to Gauss's work and to that of his ill-fated young disciple in number theory, Gotthold Eisenstein, who died at twenty-nine.

"Three steps are involved, if we are to set Bell's results in his dissertation in historical perspective," Durst explains. "First, there is Gauss's determination of the regular polygons constructible by straightedge and compass alone, as described in the seventh section of his *Disquisitiones Arithmeticæ*. Second, there is Eisenstein's further development

GOTTHOLD EISENSTEIN "one of my heroes"

and ingenious extension of Gauss's opening attack on this question. Third, there are Bell's results, which apparently carry Eisenstein's ideas as far as they can go."

In short, Bell proposed to extend the work of Eisenstein on the composition of ternary forms by raising the stakes from three to the next prime, five. This made everything more complicated, for Bell's quinary quintic is an equation of the fifth degree in five variables.

Noting the word "apparently" in the last sentence of Durst's summary, I ask, "Does Bell actually succeed in extending Eisenstein's approach to the quinary quintic?"

"I honestly can't tell," he says, "but apparently he convinced someone at Columbia that he had."

In spite of the cooperation of the late Professor Ellis Kolchin of Columbia, I am never able to find out the name of the faculty member at Columbia who approved Bell's dissertation. Since his topic was in number theory, his adviser may have been F. N. Cole, who had an interest in that subject, although his primary interest was in groups. Bell did not include in his dissertation the customary expression of thanks to an adviser—maybe because he had completed the work on his own before coming to Columbia, and felt no need. But in a 1921 paper he did express "the author's" indebtedness to Professor Cole, "not only in this paper, but for much of [the author's] other mathematical work."

Bell never returned to the subject of cyclotomy, and in only one of his later papers did he cite his dissertation; but he continued to retain an affection for the subject. In 1924, reviewing reports on algebraic numbers issued by the National Research Council, he noted with pleasure the comparatively great amount of space that the authors had devoted to cyclotomy, a fact that he saw as an encouragement to beginners and proof that the "lusty" old subject was still very much alive:

"Indeed several of the reports hint at unexhausted lodes where a novice might yet dig out nuggets of value. ... [Such] smaller finds are valuable chiefly as whetters of the appetite for discovery. The mother lode lies elsewhere."

In June 1912, just ten months after his arrival in New York, he was awarded a Ph.D. by the Columbia mathematical faculty and offered an instructorship in mathematics. It had been a long journey. Ten years after he had arrived at Stanford University, confident that he could "make it" in mathematics, he was at last formally recognized as a mathematician and, at the beginning of September 1912, a member of the mathematical faculty at one of the most distinguished universities in the United States.

In that month New York City was probably, as is usual, at its best. The muggy summer heat that drove wives and children to the seashore was gone. Although there may have been a little nip in the air, the slosh and dirt of the eastern winter lay far ahead. Bell, settled with his wife in their single room on West 124th Street,* was organizing his teaching for the coming academic year and putting the finishing touches on a paper he intended to present in October at the annual meeting of the American Mathematical Society.

The American Mathematical Society had begun at Columbia University in 1888 as the New York Mathematical Society and, although the name had changed, the annual meetings were still held in New York. At the October meeting Bell would have the opportunity, probably for the first time, to come in contact with a number of men of his own age who had already begun to establish themselves in American mathematics: George D. Birkhoff, born 1884; Oswald Veblen, 1882; Solomon Lefschetz, 1884; H. S. Vandiver, 1882, to name a few.

If he still trailed them in professional status, he had at least discovered during his year at Columbia what would be for him, mathematically, "the mother lode." The first gold was in the paper he planned to present in October—"Liouville's theorems on certain numerical functions." The immediate stimulus for this work (which, according to Bell, he had begun in 1911, possibly even while in Yreka) had been a number of unproved theorems concerning numerical functions that had been published by Joseph Liouville beginning in 1857. Liouville was a mathematician with a tremendous range of interests, but he devoted the last years of his active mathematical life almost entirely to number theory. Taking advantage of his prerogative as editor and founder of the *Journal de Mathématiques Pures et Appliquées,* he published, without any indication of the method by which he had obtained them, theorem after theorem—perhaps, Bell later suggested, an effort on his part "to emulate his immortal compatriot Fermat."

Remarkably accurate conjectures, based on a little computation with paper and pencil, are common in the theory of numbers where, as the English mathematician G. H. Hardy once pointed out, "it is sin-

---

* A "friend" in Yreka to whom Bell had lent $600 had failed to repay his debt so the young couple, with less money than they had planned on, made up the difference by living in one cheap room all the time they were in New York.

gularly easy to speculate, though often terribly difficult to prove; and it is only the proof that counts." In Liouville's case, however, it seemed clear to Bell, as to other mathematicians, that he, like his immortal compatriot, was *not* speculating.

"His theorems are of such a character that it is extremely unlikely that they were inferred by mere guessing or by hazardous induction from numerical examples," Bell noted. "This will probably be admitted by anyone who undertakes to verify any of the theorems numerically. Obviously there is a concealed rationale underlying the statements of the theorems, and indeed Liouville himself repeatedly asserted as much."

Clearly, in Bell's view, the theorems had not been found directly and individually but "by a simple and uniform method" that he, Bell, proposed to set forth in the paper he would present at the meeting of the American Mathematical Society in October.

As it turned out, this paper of his was to be presented *in absentia.*

September had scarcely begun when Bell received a wire from Professor Moritz, still the head of the Mathematics-Astronomy Department at the University of Washington. There was a place for him in Seattle. Would he accept?

Would he! Moritz offered $1200 a year. Columbia had quite probably offered more. The University of Washington was a teaching school, still without any real graduate program. Since 1900 Columbia had awarded more Ph.D.'s than any other university in the United States. But Bell did not hesitate. Classes had not yet begun. He approached Keyser and the Dean with a request to be released, and they agreed to let him go. On September 5, 1912, he wired his acceptance to Moritz.

And so Bell and his wife returned to the West, which was always to be their choice for the long term. This is what I would have written if I had not found the following item in the Siskiyou County High School Yearbook for 1913:

> Mr. E. T. Bell is teaching at the University of Washington in Seattle. Mrs. Bell is attending the Columbia University.

Up to the time I receive this information I have assumed that while Bell was getting his Ph.D. his wife was taking care of their modest living quarters, somehow managing to prepare meals in one room, visiting museums and art galleries, typing her husband's English assignments. Suddenly this view of Mrs. Bell and her activities is no longer tenable.

JESSIE "Toby" BELL studying at Teachers College, Columbia, 1911–1913

Attending the Columbia University.

But Columbia has no record of a Jessie Lillian Bell in attendance from 1911–1913, nor is there a record of her at Barnard, the women's college affiliated with Columbia. As a last resort I turn to Teachers College, not having realized earlier that its records would be separate from those of the University, and receive the following summary:

Ms. Bell was in attendance at Teachers College in the Academic Year 1911–1912 during which time she earned 34 points as an undergraduate. During the Academic Year 1912–1913 she completed 32 points as an undergraduate. On June 4, 1913, she was granted the "Specialist Diploma in Fine Arts."

Suddenly I have to revise my picture of Mrs. Bell.

When her husband decided that he preferred the University of Washington to Columbia University, she was in the middle of a course of study that would provide her at the end of another year with a diploma from Teachers College.

At this point she did exactly what any number of the "spunky" young women who were to be the heroines of the novels of "John Taine" would have done. She elected to remain at Columbia, alone.

Among Bell's memorabilia there are no letters at all from his wife. There is only a cracked and frayed photograph of himself, on the back of which he has written, "Toby kept this while I was off in Seattle."

THREE

imagination's other place

The Poetry of Mathematics

It is difficult to explain Bell's decision to leave an instructorship at Columbia University for one at the University of Washington. Having just received his Ph.D., he chose to position himself on the opposite side of the continent from the prestigious eastern universities and half a continent away from Chicago and L. E. Dickson, the leading American mathematician in his field. One cannot help wondering whether, even after receiving his doctorate, he had any thought of a career as a professional mathematician as opposed to that of a college teacher of mathematics. Was he still thinking merely of a job where he could continue to pursue his interests in mathematics and poetry, but a job at a higher level and with better pay and status than that of a high school teacher? The only explanation that Bell himself ever gave for choosing Washington is contained in a letter to Professor Smith, apologizing for not having said goodbye before leaving Columbia.

"It seemed best to come here, for the climate & experience if nothing else."

The University of Washington to which he returned was physically quite different from the university he had left four years earlier. Enrollment had increased at an average rate of fifteen percent a year, and an extensive complex of buildings from the Alaska-Yukon-Pacific Exposition of 1909 had largely replaced the forested campus. But *academically*—although President Thomas Kane continued to emphasize the need to develop graduate work in fields "in which work of good quality could be done," not much had changed. A scholarly series of publications had been inaugurated the previous year, but to date papers had appeared only in botany and English.

Special Collections Division, University of Washington Libraries, Neg. #UW14597

R. E. MORITZ, longtime head of the Washington mathematics department

The single established mathematician was the chairman of the Mathematics-Astronomy Department, R. E. Moritz. Although Moritz had come to the United States when he was twelve, he remained "very Germanic," I am told by his nephew, H. A. Arnold. While working his way through Hastings College as a carpenter, he had managed to play football, captain the team in his senior year, and win the intercollegiate long distance kick as well as the mile run. After taking a Ph.D. at the University of Nebraska and a second Ph.D. at the University of Strassburg, he had gone on for further study in Göttingen and Paris. During the years Bell was to be associated with him at Washington, in addition to publishing some thirty mathematical research papers and several mathematical textbooks, he edited a book of quotations on mathematics, wrote a scholarly paper on "The Variation and Functional Relation of Certain Sentence-Constants in Standard Literature," and invented an instrument, the cyclo-harmonograph, for drawing large classes of the higher curves. To Bell he was always to be "the Pterodactyl."

Moritz "used a good deal of influence to keep his men abreast of the times," according to President Kane, but as chairman he had to accommodate a large and increasing number of elementary classes with a relatively small faculty. It is not possible to ascertain all the classes Bell taught his first year. Certainly his teaching load was heavy—he complained of twenty-two hours a week—but he is listed specifically in the catalogue only as instructor for Theory of Numbers ("an introductory course") and Theory of Functions of a Complex Variable.

I am able to locate only two letters from Bell relating to this first postdoctoral year. Both were written in the fall of 1912 to Professor Smith back at Columbia. In the first Bell expressed his regret at leaving the "fine libraries" of the East—in particular the *Arithmetica* of Diophantus with the notes by Fermat.

"[But] after having thought about numbers for another year or two, I should be better prepared to appreciate what Fermat leaves unexpressed. The man who successfully summarizes Fermat's work in Numbers will have to possess a more detailed knowledge and feeling of the spirit of Diophantine Analysis than has anyone who is now writing." A proper evaluation of Fermat's work could be made, in his opinion, only by someone *"into whose very bones the spirit of discrete, as opposed to continuous, quantity is bred* [my italics]." Edouard Lucas, whose *Théorie des Nombres* he was again studying, had been the only mathematician in the nineteenth century "who had the right spirit." He had obtained results by his methods "that only come with great labor by the usual means." (Here Bell was referring to Lucas's "symbolic method," known today as the umbral calculus—a tool that Bell himself was later to utilize with great effectiveness.) Lucas's "elementary and apparently simple analysis," Bell wrote, was "often more profound and germane to arithmetic than all the methodical algebra of Gauss & his successors." But, he lamented, "the older intuitive ways of looking at things are not in fashion at present. Even Lucas is sometimes sneered at for his elementary way of going at things, and it is only recently that people have had the courage to quote freely from his work."

Along with his intensive study of Lucas, and in spite of the heavy teaching load and the absence of his wife, Bell found time to continue the mathematical research he had begun before he left Columbia. He does not seem to have been in any great hurry to publish his results, or perhaps he did not really know how to go about doing so. Instead he contributed a number of problems to the Problems and Discussion Section of the *American Mathematical Monthly,* a practice he had begun even before leaving Columbia. There were three problems in 1913, six in 1914, two in 1915, four in 1916. He also continued to write poetry.

Mathematician as poet, or poet as mathematician—these were not callings that were antithetical in Bell's view. Many years later, reviewing a collection of essays by Professor Smith, titled *The Poetry of Mathematics,* he subscribed wholeheartedly to the tenet Smith put forth, "a thesis [Bell's words] which may be succinctly stated by saying that mathematics and poetry are simply isomorphic ... both the mathematician and the poet are creators."

In the second of his letters to Professor Smith, Bell went back to the subject of Lucas and to his concern about the nuggets that might remain in the unpublished work of that mathematician, who had died at forty-nine, never having completed the promised second volume of his *Théorie des Nombres.* In a roundabout way, he encouraged Smith, as someone of greater mathematical stature than himself, to inquire about the unpublished work. Most tantalizing to Bell was the fact that at Columbia he had happened across a note in which C. A. Laisant had recalled an almost complete proof of Fermat's Last Theorem that Lucas had once described to him. At the time the proof had sounded quite reasonable to Laisant, but later he found that he had completely forgotten it! At Bell's urging Smith wrote to learn more. Laisant replied —Smith sent the letter on to Bell—that unfortunately there was nothing about Fermat's theorem in Lucas's unpublished papers. For the next forty years, however, the possibility of material left by Lucas for the promised second part of the *Théorie des Nombres* tantalized Bell, and other number theorists.*

From the heading of the above letter to Smith—University of Washington—and the date—December 27, 1912—it is clear that Bell did not go to New York during the Christmas holidays to be with Toby. Whether she came to Seattle is not so clear, since there is no mention of her in the letter. What is certain is that she remained at Columbia until June, when she was awarded her "Specialist Diploma in Fine Arts."

It occurs to me that with her new credential she might have obtained a position in the Seattle schools during the coming year; but before I can investigate that possibility, I happen to come across her name in the faculty-student directory for 1913–1914. Since faculty wives are never listed in the directory, it is clear that her name is there because she was a student. I expect to find her continuing her education with courses related to art, but such was not at all what Toby Bell had had in mind. She was majoring in mathematics.

I can say very little about the Bells' personal life in these early days at Washington, other than that they both worked very hard and kept on the whole to themselves, "selecting their own society." Toby, in addition to doing the assignments for her classes, did whatever typing was required in the family. She also worked on the woodcuts for the book in which *The Scarlet Night* was to be copied.† Bell prepared for his classes and continued his study of Lucas and the research that he had begun at Columbia in connection with Liouville's theorems. He may have worked as well on other specific problems in number theory, but he published nothing. In the spring of 1914 he embarked on the first

---

* Lucas had died at forty-nine as the result of a freak accident, a waiter at a banquet having dropped a tray of dishes on his head.

By the time Laisant wrote to Smith, Lucas's number theoretical manuscripts had already been passed on to André Gérardin, the editor of Sphynx- Œdipe. Becoming increasingly eccentric, Gérardin would permit visiting mathematicians to gaze upon the crate of "Lucas's papers" in his one-room apartment but never to see the papers themselves. At one time he was asking $30,000 for the treasure. "In all his life Lucas never had that much money," Bell groused. When, however, Gérardin died in 1953, it was discovered that the crate had contained merely copies of published papers. The question of what happened to the number theoretical manuscripts of Edouard Lucas remains, to this day, unanswered.

† The completion of the woodcuts is recorded in her hand on the flyleaf of the volume —"(1-1-1917/12 p.m./Fin. all)." Although she continued to work on the project throughout her life, she never finished all the coloring and illumination, and the poem was never copied in the book as planned.

substantial narrative in verse that he had attempted since *The Scarlet Night*. He called it *The Enlightened Doctor*.

That same spring the San Francisco Section held a meeting in Seattle, an unusual occurrence and one that permitted Bell and other mathematicians at Washington to present their work in person. In a third letter to Professor Smith, dated a week before the meeting, Bell wrote, "Work here is plentiful, if nothing else, but I have managed to get a lot (as far as bulk at any rate is concerned) done, & will try to get some of it published if the opportunity offers."

Since Bell was still not a member of the American Mathematical Society (again, curious for someone setting out on a career as a professional mathematician), he had to be introduced to the Section by someone who was a member, in his case another young instructor at Washington. The more important of the two memoirs offered by Bell was titled "An arithmetical theory of certain numerical functions" and was a continuation of the work begun at Columbia on the Liouville theorems. The paper is the first real example of what was to be Bell's characteristic approach to the theory of numbers. As he saw it, then and always, the theory of numbers—Gauss's "Queen of Mathematics"—was anything but in control of her subject. Number theory was "an unruly domain . . . the last great uncivilized continent of mathematics . . . split up into innumerable countries, fertile enough in themselves, but all more or less indifferent to one another's welfare and without a vestige of a central, intelligent government. If any young Alexander is weeping for a new world to conquer, it lies before him."

Even on a lesser scale than world conquest, the theory of numbers was singularly lacking in general theories. It was for the most part simply a collection of special results and methods. Throughout his mathematical career Bell was to strive to develop general methods and theories from which large numbers of previously disconnected, individual number theoretical results could be directly obtained. The work he presented to the meeting in Seattle in the spring of 1914 did exactly that.

His first step, while still back at Columbia, had been to discover "the simple and uniform method" by which Liouville had actually proved the theorems he had merely stated in his journal.

"The fact is that Bell had figured out what Liouville was up to and how he got his results using generating functions," Durst points out. "Thus it was as easy for him as it had been for Liouville to crank out theorems whose proofs heretofore had required an Euler or a Gauss or that powerful pair together."

Bell's primary goal in his paper, however, was "to construct a self-contained *arithmetical theory* of a large and important class of numerical functions, the theory to be so formed that the interrelations of the functions considered . . . shall be exhibited with a minimum of calculation, from either their symbolic or verbal definitions." A numerical function assigns to each integer some quantity that is determined by that integer—for example, the square of the integer, the number of divisors or the sum of the divisors, and so on. What Bell did was to create an algebraic system in which the elements were not the ordinary numbers but, instead, the numerical functions themselves. The expression *arithmetical theory*, as he used it here and later, refers to an algebraic system in which there is unique factorization, as in the natural numbers $1, 2, 3, \ldots$*

"This is the paper of Bell's that I like best," Tom M. Apostol of the California Institute of Technology tells me. "In it Bell proved many important things about numbers that he had discovered; but because the paper was published in a little known series, later mathematicians rediscovered the same things and—unaware of Bell's work—proved them again. He introduced among other interesting things the idea of regarding certain arithmetical sums (now known as Dirichlet convolutions) as a kind of multiplication with interesting algebraic properties. I think it is a *very* interesting paper."

As Apostol points out, Bell's choice for publication of his first paper was unfortunate. Instead of sending it to one of the established mathematical journals, he chose to make it the initial paper in the *University of Washington Publications in Mathematical and Physical Sciences*. In this choice he exhibited a certain professional naiveté—if he was interested in making a mathematical reputation for himself. Although publication at Washington had the advantage of being faster than publication in a journal and perhaps carried with it a certain immediate prestige among his colleagues at Washington, it had the disadvantage that his paper would appear without corrections and suggestions that a qualified referee might have made. In fact, it has been suggested that

---

* The theorem that states that every number is uniquely expressed as the product of primes is known as the Fundamental Theorem of Arithmetic. Throughout Bell's mathematical career he was always interested in discovering analogues of the Fundamental Theorem in areas other than the integers.

Bell may have selected the in-house series just so that his paper would not be refereed. Perhaps, though, he chose the Washington series to add mathematics to university publications that had heretofore been limited to other fields—or simply because, as he indicated in his letter to Smith, "the opportunity offered."

Whatever his reason, the result was that the principal review journal in mathematics did not summarize the contents of his paper but noted its appearance only by title.

After Durst begins to organize and evaluate Bell's mathematical papers, I ask him to select a few—say, ten percent, or some twenty papers—that he considers of special interest for various reasons. He responds with an annotated and sometimes starred list that he describes deprecatingly as "probably suffering from an excess of subjectivity on my part, due in no small measure to eccentricity and various shortcomings." In it he emphasizes some of Bell's central topics and indicates a few of their consequences. Although he objects that this is an approach "somewhat analogous to studying comets by examining their heads and all but ignoring their tails," it seems to be all that is possible in a nontechnical narrative that aims to present the man rather than his scientific work.

Of the University of Washington paper (to which he gives two stars out of a possible three), Durst writes: "In this work Bell extends and organizes a large number of results of Liouville ... by constructing the ring of number theoretical functions, as some choose to call it these days. He also develops a technique using what have been referred to by Tom Apostol as 'Bell series' in order to derive in a simple and straightforward manner numerous special results for factorable, or multiplicative, functions announced by Liouville."

Like Apostol, Durst finds it a remarkable paper and marvels that it seems "all but unknown, judging by the frequency with which parts of this work are still being 'rediscovered' by other people." But he concedes that it probably deserves a prize for obfuscation.

"The real question is just what did Bell do or not do in the paper. It's very hard to tell. He has things spread out far and wide and is constantly citing earlier sections as well as later ones (some of which don't even exist) in order to justify his claims or suggest where their proofs may be found, or where some of his remarks might be retracted or modified. The closer I studied the text, the more the word 'etc.' became a red flag. Two of my favorite instances—the following complete sentence: 'Hence, etc.' and my very favorite footnote: 'A set of elements form a semi-group if they have the *group property,* etc.' That's the end of the footnote. Why he chose to present his discoveries in such a confusing way, I have no real idea."

I suggest that perhaps he is seeing unleashed in this mathematical paper some of the exuberance that Bell was later to channel into his science fiction novels.

"You may be right," he agrees. "Obviously he needed someone with whom he could bounce ideas back and forth, someone who could give him some fatherly advice about writing mathematical papers so people could read them. In later years he certainly showed every sign of having learned a good deal about mathematical writing as well as other kinds."

At the time "An arithmetical theory . . ." appeared, Bell was also eager to see in print the poems he had been writing since his days at Stanford. Somehow he had learned of the Gorham Press in Boston. A family firm, headed by Richard G. Badger, it published various series on subjects ranging from literature, economics, and ethnology to "rational sex." Many of its publications were Ph.D. dissertations and books by scholars who could not find commercial publishers. The firm had also been publishing for the past quarter of a century a magazine called *Poet Lore.*

On April 11, 1915, Bell signed a contract with the senior Badger for a book of poems titled *Recreations,* an appropriate title since this was what—at least in later years—he always said he considered his nonmathematical writing to be. He was to pay Badger $675 to publish his book and to receive forty percent of the gross proceeds of the sales. On the back of the first page of the contract there is an addendum signed by both men:

> It is understood and agreed that said work is to be published under the pseudonym "J. T." and that the anonymity of the author shall be preserved at his discretion.

In August 1915, the same month that *Recreations* appeared, Bell signed yet another contract with Badger.* This was for a book of poems titled *The Singer.* The contract for the second book, with the same

---

* Both books were copyrighted by Richard G. Badger, a name that seemed so unlikely to James Rock, the editor of a compilation of pseudonyms of science fiction writers, that he dismissed it as probably yet another of Bell's pseudonyms. As a result it has been so listed by a number of writers on "John Taine."

financial terms, also contained the agreement as to the anonymity of the author. The degree of monetary risk to Bell and his wife in these ventures can be gauged by the fact that their payments to Badger were the equivalent of a full year of Bell's salary.

Bell immediately sent off a copy of *Recreations* to Professor Smith. The latter was something of a poet himself, having translated the *Rubáiyát of Omar Khayyám*—yet another mathematician who wrote poetry. *Recreations* was, Bell explained to Smith, "a curiosity showing what the mathematically minded do in their spare time." It was a long way from mathematics, strictly speaking, "but many of the better things in it have great significance if read by one who is not unacquainted with science in general, and mathematics in particular." He called Smith's attention to poems about Newton, Clerk-Maxwell, and Lord Kelvin as well as one titled "The King," which he explained referred to Clifford. "These at least give a new kind of devil his due,—in poor form, it is true,—in general literature.

"There is more (and worse!) to follow ... ," he warned, for *The Singer* was by then with the publisher. "And I hope soon to get some really great publishing house to take a work of a more substantial character,—this last does amount to something."

Bell's two privately published volumes of verse are quite different in nature. *Recreations,* consisting for the most part of earlier efforts, is a collection of conventional poems. One reads it with the rhythms and words of the Victorians in one's ears. In a dedicatory verse the poet ponders what his "far-off song" can "waft" to readers from "youth or maiden" to "aged and lonely." Most of the poems are short and grouped under such headings as "Out of the Past," "One Day," "Beside the Sea," "Seeking."

The voice of the second volume (written in April 1914) is recognizably that of E. T. Bell as he was later to be known—extravagant, outrageous, unable to resist an opportunity to shock or to puncture pretentiousness or pomposity. Most of the poems in *The Singer,* including the title poem, are lampoons. The volume ends with a satirical "oration" poking fun at academic conservatism as exemplified by Phi Beta Kappa—"Our Motto reads: 'The best has All been said ... .'" Since the book, in Bell's opinion, was "warm enough to heat several kettles," he asked Smith not to mention that he wrote as "J.T."

"If it got out here,—especially when some of the later things see print,—that I do this kind of business, my job would not be worth five minutes' purchase,—and unfortunately I must live."

In this letter to Smith he mentioned as well, "At present I am having a good time with the theory of numbers, & seem to have got out something of value." The letter is undated, but since *Recreations* appeared in 1915, he was clearly referring to the paper that had appeared at the University of Washington in August of that year. But he did not send or even offer to send Professor Smith a copy of his first paper since his Ph.D.

I comment to Taine, "The thought has passed my mind that your father might have preferred being a poet to being a mathematician."

"I think he might have," Taine replies, "if he had had any success."

JOSEPH LIOUVILLE: "I would like [his picture], for personal reasons".

Metamorphosis

"There is no test, so far as I know, of a man's ability to be a creative mathematician . . . except that of creation," Bell wrote in his 1931 letter to the *Eagle*. "Not until [he] is thrown out on his own in actual working conditions can it be said whether or not he will produce, and the one test is *production of scientific work*. If this is not done either before the Ph.D. or in the first five years after taking the degree, it almost certainly never will be." The "brutal verdict" of the statistics was that these first five years spelled success or failure. And failure would "probably be for life."

After reading these strong words, I turn back to Bell's bibliography to see what he himself produced in the five academic years that followed his Ph.D. It is almost always difficult to determine just when Bell did his research. Sometimes he put a date at the end of a paper. Sometimes a journal noted the date on which a paper was received. (Today this is customary but not then.) For the most part one must rely upon the month and year of publication. On the basis of such dating Bell produced only one piece of mathematical work in the five years that followed his own Ph.D.—the paper published in 1915 in the *University of Washington Publications in Mathematical and Physical Sciences*.

I can offer no reason for this long, seemingly unproductive period at the time when, according to mathematical experience as well as to legend, a mathematician does his best, most original work. I can only describe the situation in which Bell was working and what the record shows that he was doing.

"Those first five years," he wrote, "often—as a rule, in fact—will be spent largely in uncongenial drudgery, far from scientific centres and frequently without ... adequate libraries. A man must adapt himself to what he finds, and he must make up his mind to work at his own science in his spare time—usually at night. He will work hard, often inhumanly hard ...."

The paper that Bell published at Washington in 1915 was an important paper with many original ideas, but it went essentially unnoticed. Perhaps discouraged—for his dissertation had had very much the same reception—he did not produce the promised "Part Two" for a number of years. (He never explained the delay other than to say that "circumstances prevented.") He still did not join the American Mathematical Society, although other young mathematics instructors on the faculty were members. (He did, however, join the new Mathematical Association of America, which had been formed to promote teaching as well as research on the college and university level.) He did not submit any more papers to the twice yearly meetings of the San Francisco Section. My suspicion is that he continued to study and work in mathematics very much as he had before he went to Columbia, but all the published record shows is that he was writing poetry. *Recreations* had appeared in 1915, and *The Singer* followed in 1916. The first of these carried a wry epigraph from Robert Browning that may have been appropriate: "So I said 'To do little is bad, to do nothing is worse'—and made verse."

When one writes of these years, however, one must remember that they spanned a period of worldwide turmoil. England, France, and Russia—along with the smaller countries that formed "the Grand Alliance"—had gone to war against Germany and Austria-Hungary in August 1914. In the following month Germany had shocked the world by invading tiny neutral Belgium. Although Bell shared the general horror and disgust at the reports of atrocities committed by "the Huns" (one of the poems in *The Singer* is "William in Belgium"), there is nothing in his admittedly few letters that indicates a concern for England at this time, or expresses the hope that the United States will come to her aid. Nor is there any indication that, although he was still a British citizen, he considered returning to England for military service, as did his brother.

The only existing letter from his mother comes from the years before the United States entered the war. In it she thanked him for sending a pound, the equivalent of five American dollars, to Enid ("She was so pleased. So was I. Good kind Hoggie.") and reported that "Flapps" was in a training camp, "grooming their horses and doing

Special Collections Division, University of Washington Libraries, Neg. #UW3831

HENRY SUZZALLO, president at the University of Washington 1915–1926

such chores." James Redward Bell went later to the front, where he performed the hazardous duties of a courier for a cavalry unit.

On the campus of the University of Washington in the early years of the World War there was also turmoil of sorts. A newly elected governor and his newly appointed Board of Regents had agreed that it was time for President Kane to go; at the end of the year 1913–1914 he was informed: "The University has grown beyond you." The Regents then conducted a search for a president "capable of doing for the University of Washington what Dr. Northrup did for Minnesota, what Eliot did for Harvard—what Butler has done for Columbia and what Hadley is doing for Yale." They found their man in Henry Suzzallo of Teachers College, Columbia University, who took over as president in

September 1915. He was welcomed as a man who "combined in his outlook the virtues of a western man with eastern experience."

Bell had quite a bit in common with the new president, but it is unlikely he ever mentioned anything to that effect. Although Suzzallo was seven years older, he also had grown up in San Jose, also had taken an A.B. at Stanford. He was in fact an instructor "on leave" when Bell was at Stanford.

At the time of Suzzallo's arrival the mathematical faculty consisted of one professor, two associate professors, three assistant professors, and five instructors. Among the instructors Bell's name was first in the catalogue although the list was neither alphabetical nor according to seniority. His teaching load and the classes he taught were much the same as they had been during his first year at Washington. His introductory course in the Theory of Numbers was not regularly offered, and in 1915 he took on a subject that was surprising in view of his long interest in number theory—the Applications of Mathematics to Physics.

In addition to his teaching, there is only one thing I can say for certain that Bell was doing. He was writing poetry. Reams of poetry. And trying very hard to get it published commercially. Typescript copies of poems written at this time, sent out to publishers and regularly returned, are among the unpublished Bell papers that Taine gave to the University of California at Santa Cruz. They include a number of short poems and a few fairly long ones. Among the latter are "A California Valley," "The Devil's Funeral," and "The Prince," a revised version of "The Singer." The longest poem is *The Enlightened Doctor,* which is booklength and which like many of his science fiction novels deals with the legend of the Philosopher's Stone. In fact, he considered the section titled "Gold" the best and most important. In the poem the dying Ramon of Palma (the enlightened doctor) goes back over his life, describing his love for Bianca and for his horse Typhoesus, and his efforts to cure the cancer that Bianca develops after the jealous horse kicks her in the breast. The love story, which begins in childhood, has a tenderness that is absent in Bell's novels, and there is a hint that it may have had roots in his childhood in San Jose.

Ramon ranges widely over the meaning of life and death; but in the end he finds, written in gold in the sealed pages of the Book of Knowledge that his godfather gave him on his fourth birthday, the words *Fool's Gold.*

In 1948, when Bell was revising *The Enlightened Doctor*, still with the hope of publication, he noted on the typescript that it was written in Seattle in February 1916. He retained as well Toby's original typescript, which carries the dedication:

To
CASSIUS JACKSON KEYSER
Poet, Philosopher and Mathematician
a tribute of friendship and esteem

In a "Note to the Reader" Bell explained that one of the central ideas of the enlightened (or "illuminated") doctor had come from an address by Keyser on "The Significance of Death."

The note was signed JOHN TAINE.

This is—as far as I have been able to ascertain—Bell's first use of the pseudonym John Taine. Although "By John Taine" appears on the typescript of "A California Valley," it is there as a handwritten insertion, not as a part of the typescript as in the case of *The Enlightened Doctor*.

Bell's son, Taine, has always supposed that his father took the name from that of the French philosopher-writer Hippolyte Taine (1828–1893) and thinks that he recalls his father telling him that it was "the name of a French philosopher." There are no books by the philosopher Taine among those that Bell still had at the time of his death; however, his *Notes on the English*, a study in the style of de Tocqueville, was reissued in English a number of times during Bell's youth. It is a book that would have appealed to him, for many of the author's comments—particularly on the English class system—echo his own comments on the subject.* (In later years, when a woman friend asked Bell how he happened to take the name "John Taine," he replied that it was "a family name." But I never come across it on either side.)

The Enlightened Doctor again treats the legend of the Philosopher's Stone, which had attracted Bell at least as early as 1910, when he wrote *The Scarlet Night*. The Stone was reputed, not only to transmute gold from baser elements, but also to abolish death itself. Although Bell dealt with both aspects in his poem, Professor Keyser had treated only the latter, putting forth the idea that death gives significance to life.

---

* Bell always identified himself as a Scot and customarily distinguished between mathematicians born in England and those born in Scotland, such as the Scottish algebraist J. H. M. Wedderburn, whose name he gave to a character in *The Purple Sapphire*.

The writing of *The Enlightened Doctor* followed very closely upon the twentieth anniversary of the death of Bell's father, a date he had begun to memorialize the year before in the Caesar. Its conception may have coincided with that anniversary, for much of the poem focuses on the subject of death and also on the childhood recollections of Ramon, again at the age of four.

> Four years old.
> How ancient then I seemed, how wise and bold;
> They spread a feast, made much of me that day.
> Death grasping all shall not take this away.

A short poem, "Before Sleep," which is in the same folder with "A California Valley," echoes the lines.

> Before sleep visits me
> In one swift flash I see
> Myself when four years old
>
> Again I live in him
> Whom life shall never dim
> Nor any chance unmake

In his 1948 revision of *The Enlightened Doctor*, Bell rewrote the note to the reader and omitted the reference to Keyser, who had died the year before: "The dream of the alchemists was realized in the twentieth century [he was by then writing three years after the creation of the atom bomb], but in no such shape as Ramon anticipated. The darker mysteries that colored his thoughts and actions still shadow us. In this sense he and we are contemporaries . . . ."

In *The Enlightened Doctor* he also had something to say about the relationship between a father and his son.

> Should you
> Some day have a boy to call your own,
> Though he be your own flesh and your own bone,
> Think not thereby you hold his nature's clue.
> As two concentric circles never touch,
> You shall not know him though you love him much:
> Dividing at the fount, your lives shall be
> Separate rivers to eternity . . .

It was, coincidentally, at the beginning of the following year (1917) that Bell learned that he was to become a father. As Taine has told me earlier, neither of his parents welcomed the news.

Toby was well into her pregnancy that spring as the United States entered the war in Europe on the side of the Grand Alliance. In June she received her B.S. from the University of Washington. I find the transcript of her record interesting. In addition to meeting such requirements as political science and foreign languages—she took two years of French and two of Italian—there is an unexpected semester of Sanskrit. Also interesting, because of Bell's later enthusiasm for collecting Chinese art, is a year course in Oriental History and Literature. In her major subject of mathematics, except for a B in Vector Analysis in her last year, there are straight A's.

When classes began in September 1917, although the University was still offering most of its usual programs, enrollment was down thirty percent because of the war. Bell, who had been promoted to assistant professor the previous September, was teaching classes composed largely of women.

Just a few weeks after the beginning of the semester, on September 27, 1917, Toby delivered a healthy baby boy, who was given the name Taine Temple Bell.

The more I learn about Bell's family background the more I find it surprising that Bell did not call his son James. For at least four generations, going back to the beginning of the nineteenth century when a certain James Bell fathered a son down by the Billingsgate Fish Market, James had been the traditional male name in Bell's family. His great-grandfather, his grandfather, his father, and his older brother—the first son had always been named James. It seems odd to me that Bell, who had taken James Temple as his first nom de plume, had not given the name to his son. Ultimately I happen across Taine's birth certificate. On it the infant's name is written *James Tain Temple Bell*. Then someone has added an *e* to *Tain* and crossed out *James*.

At the beginning of the year in which Taine was born, but even before Bell knew that he was going to be a father, a change seems to have occurred in his attitude toward mathematics as a career. For the first time, as of January 1, 1917, his name appeared on the membership list of the American Mathematical Society. In August 1917, two years to the month since his first and only paper published at Washington, he had another paper, a short one, in print—this time in an English journal, the *Messenger of Mathematics*. Two more short papers had also been accepted for publication and were to appear in the first half of 1918. He sent brief reports on these to the Berkeley meeting of the San

Francisco Section. Since he was not present, they were read merely by title, but abstracts appeared in the report of the meeting.

While his mathematical work was thus beginning to be published, it was otherwise with his long poems, which continued to be regularly rejected. A letter from the Gorham Press informed him that sales of *Recreations* and *The Singer* had been so poor that the Press had decided to remainder both books. He could purchase as many as he wanted—for thirty cents apiece—the rest would be destroyed. He ordered twenty-five copies of each book. Today a California dealer lists *The Singer* at $3500 but states incorrectly that no other copies exist.

Bell was always to be bitter about his experience with the Gorham Press. He had expected his books not only to be printed but also to be publicized. Instead, as he saw it, the Press had taken his money, published his books, and then offered to sell them back to him. He took his poetry seriously. When, a few months later, he received a formal letter returning the copyrights to him, he marked it "important" and "to be kept."

There were to be no more published poems, but there was to be new and extensive mathematical research made public, although not yet being published. In October 1918, a month before the World War ended, Bell sent five papers to the meeting of the San Francisco Section at Stanford. Four were subsumed under the title "Arithmetical Paraphrases I, II, III, IV." These were eventually to form the key paper in the second and most extensive area in which Bell worked.

"Bell seemed to have a real talent for selecting problems whose resolution made it possible to produce a multitude of special results of interest and importance in number theory," Durst points out.

In each of the principal areas of his research, as Durst classifies them, there was one main theme that he subsequently exploited. In the first—Arithmetical Functions and Theories—the theme was set forth in "An arithmetical theory of certain numerical functions." In the second—Forms, Elliptic Functions, and Paraphrases—it was laid out in "Arithmetical Paraphrases."

In this ambitious work Bell had gone back once more to Liouville's journal. There, between 1858 and 1865, in eighteen scattered memoirs, the French mathematician had published, again without any indication of how he had obtained them, another group of remarkable theorems under the modest title "On some general formulas that can be useful in the theory of numbers." Liouville and others had applied these formulas to various questions of interest, but the fact that no one knew how Liouville had managed to obtain them—as Bell later wrote to the *Eagle*—"had worried mathematicians for some time."

Gleanings from abstracts (presumably submitted by the author) can convey to the specialist something of the nature of the work and to others at least a sense of its significance:

"By means of the theorems in his first paper and another simple principle, Professor Bell shows that the classical expansions in the theory of elliptic and theta functions may be paraphrased directly into theorems of the Liouville kind [and] that a simple method is thereby provided for the discovery at will of new results of the same general sort.

"In his second paper [he] prepares and collects elliptic and theta expansions in a form suitable for paraphrase, and gives sixteen trigonometric identities basic for Liouville's numerous theorems on the quadratic forms of certain primes. . . .

"In his third paper [he] applies the methods of his first to a few of the series in the second. . . .

"His fourth paper applies the method to the series for non-doubly periodic theta quotients given by Hermite (1862) and extended by G. Humbert (1907) . . . ."

The work on paraphrases, like the earlier work on numerical functions, bears the stamp of Bell's mathematical personality, according to R. P. Dilworth, a later colleague.

"There were a lot of number theorists at this time who were not specifically interested in this kind of [general] approach to number theory but rather were interested in a range of problems to which paraphrase has applications. The fact that Bell came up with this idea, exploited it, and then worked so hard on developing it was, I think, branching out in a quite different way from what other number theorists were doing at the time. It reflects a little of his tendency to be different and original."

"Arithmetical Paraphrases" was not to appear in print for more than two years. Its significance, however, was apparently recognized right away by Moritz, who passed on the word to Washington colleagues outside mathematics.

The next year the botanist T. C. Frye suggested that a "research society" should be formed consisting of faculty "who [had] gained a national reputation for productive scholarship through ten years of investigation along lines of original research of high quality made public in permanent form." A "disinterested committee" from Suzzallo's of-

fice proceeded to select eleven charter members from ten departments
—mathematics, with Moritz and Bell, being the only department to
have more than one member. Since it was just seven years since Bell's
Ph.D. dissertation, an exception to the ten-year rule was obviously
made in his case on the basis of "Arithmetical Paraphrases" and other
work he was producing.

The year 1919, in which the Research Society was founded marked
a turning point in the attitude toward research at the University—and
in Bell's life as a writer and as a mathematician.

During the immediately preceding years he had been fascinated
by the idea of the Devil, although not in a religious sense. As he might
have said, he just liked the idea of the guy. His earliest "Devil" poem
was "England's Secret" (1915). He followed that with "The Devil's
Funeral: An Elegy" (1916), based on a newspaper report that the
Methodist Episcopal Church was proposing a special service for the
burial of children that would substitute *Sin* for *the Devil*. Next came *A
Weekend in Hell* (700 pages in Toby's typescript), in which he combined
prose (both narrative and drama) and verse. Conventionally framed as
a dream, *A Weekend in Hell* is better described as a hallucinatory experience brought on perhaps by the drinking of alcohol. (The Eighteenth
Amendment—"Prohibition"—did not become law until 1919.) During
the "weekend" Bell and his cat (who addresses him as "Pops Romps")
spend three days with his Satanic majesty and almost every historical
character the reader can call to mind. The work was completed at 12:05
a.m. on March 30, 1918. Bell continued to write short poems from time
to time, but in summer 1919 he turned to a completely new category
of literary expression: the writing of scientific adventure novels.

At almost the same time there was an even more startling change
in regard to his mathematics. During the academic year 1919–1920 he
published almost twice as many papers as he had published in the
seven years that had passed since his Ph.D.—

In September 1919: "The twelve elliptic functions related to sixteen doubly periodic functions of the second kind."

In October 1919: "On the number of representations of $2n$ as a
sum of $2r$ squares."

In December 1919: "Sur les representations propres par quelques
formes quadratiques de Liouville."

In January 1920: "On class number relations and certain remarkable sums relating to 5 and 7 squares."

In February 1920: "Definition and illustrations of new arithmetical group invariants."

In March 1920: "Parametric solutions for a fundamental equation in the general theory of relativity, with a note on similar equations in dynamics" and "On the enumeration of proper and improper representations in homogeneous forms."

In April 1920: "The equation $ds^2 = dx^2 + dy^2 + dz^2$."

In May 1920: "On a certain inversion in the theory of numbers."

In July 1920: "On the representation of numbers as sums of 3, 5, 7, 9, 11 and 13 squares."

How does one explain this sudden mathematical creativity after seven seemingly fallow years following the Ph.D.?

Durst recalls that when he was a student at Caltech, Bell himself provided the following explanation. At a certain point in his career he asked a simple question (of whom he did not say). The question was *How does a man get ahead in mathematics?* He received a simple answer. *Publish.* And publish he did.

E. T. BELL with his son, Taine Temple Bell

Parallel Worlds

Self-sufficient and happy hermits.

So their young friend Glenn Hughes (later himself a professor, poet, and playwright) described the Bells during the years that followed the World War. They bought a house "on the installment plan," a modest bungalow in an area favored by young academics and people outside the university world: dentist, musician, baker, engineer, salesman. Immediate neighbors were the dentist and the engineer. But the Bells kept to themselves. They did not entertain and, of choice, had only a few friends. These were startled when they called and found the living-room and dining-room walls painted gold—"not just yellow, but gilt." One friend maintained that Bell, whose interest in the Middle Ages was well known, was "reflecting on his walls the romantic dream of the alchemists." Another suggested that Mrs. Bell "considered gold the most flattering background for herself."

The quotations above come from Hughes's *Academe,* a series of faculty sketches in which he presented Bell as a chemist named Andrew Sears. The story of the gilt walls also appears in his profile of Bell for the *Stanford Illustrated News.* Although both pieces were written later, I have used material from them at this point since the friendship dated from 1919 when Hughes came to Washington as a Denny Fellow in English—a congenial young man who submitted a volume of verse for his master's degree rather than the usual thesis.

Surrounded by his gold walls, a pipe always in his mouth or lying beside his manuscript, a cat often rubbing against his leg, Bell worked at a table in the living room—worked and worked. Taine, a toddler by

Photography Collection, HRHRC, The University of Texas at Austin

GLENN HUGHES at a later time with his wife, Babette

then, was cautioned not to disturb his father and later could not bring other children home. Even adult visitors were discouraged.

To produce in the situation in which Bell found himself, a man "[had to] be prepared to sacrifice a large part of the so-called social pleasures, and to stick strictly to business . . . to work a double shift . . . to work hard, often inhumanly hard . . . if he [was] to forge ahead and not write himself down as a failure at thirty-five or forty."

Come summer, though, he could permit himself the pleasure of "recreational writing." In the past this had been poetry but now, as "John Taine," he began to write what he called "scientific adventure yarns"—what a publisher later classified as "scientific mysteries"—and what still later came to be known as "science fiction." In his Bell profile, titled "The Master of Scientifiction" (the genre also passed through many name changes), Hughes gave precisely the year of this change.

"As a relief from the grind of mathematics, Bell began writing novels in 1919."

So—*1919*. Obviously the failure of his poetry to sell had had something to do with this abrupt change of direction. It may also be more than a coincidence that the change occurred in the same year as the founding of the Research Society, for this brought him into closer contact with faculty in other sciences.

"Always possessed of a deep interest in creative literature (both poetry and prose), and of a brain overflowing with scientific information, he let loose his imagination," Hughes continued, "and the result was a novel of gripping interest and magnificent twentieth-century fantasy."

But which novel? Hughes, who must have known, did not say.

I assume that Bell's novels were written in the order of their publication. There are four novels by Bell that were never published. Logic supports the assumption that these novels, rejected by publishers even after "John Taine" had become a well-known name in science fiction, were his earliest efforts in the new field. But logic is not to be trusted in the case of Bell. As he himself wrote in *The Time Stream*, "Whenever you find a perfectly reasonable explanation of anything in nature or human conduct, look for something else." Holograph copies of three of the four unpublished novels, all currently in the hands of dealers, are dated in Bell's hand at the time of writing, and all are dated after 1919. So, I wonder, could his first novel have been *The Purple Sapphire*, the first published novel? Logic again. But no. Reading Bell's personal copies of the novels in the order of their publication, I come in time to the fourth, *Green Fire*, published in 1928. There on the flyleaf is a note—"My first sf story Seattle June 1919."

Although *Green Fire* is subtitled "The Story of the Terrible Days in the Summer of 1990," it is of no interest as a futuristic novel. Its interest lies rather in its relation to earlier experiences in Bell's life.

The first of these was his discovery of *The Interpretation of Radium*, the little book by Frederick Soddy that had so impressed him when he first came across it in Yreka and found expressed there ideas that had been implicit, although not articulated, in his *Scarlet Night*. On page 136 of *Green Fire* the "mad scientist" Jevic reads aloud to the heroine from *The Interpretation of Radium* for four and a half printed pages. That is a lot to quote from another man's book! All is from Soddy's final chapter (dropped from subsequent editions) in which he proposed that such legends as the Fall of Man might be relics of some ancient and unsuspected civilization that had "attained not only to the knowledge we have so recently won , but also to the power that is not yet ours." Such dominance might not have lasted long. "By a single mistake, the relative position of Nature and Man as servant and master would . . . have become reversed, but with infinitely more disastrous consequences . . . ."

Such a catastrophe—at the hands of the insane Jevic, who has discovered how to release the power in the atom—does not occur in *Green Fire*, being averted by a battery of brilliant Scottish scientists. (All the "good guys" in *Green Fire* are Scots.) But Soddy's idea of the highly advanced civilization and the fatal mistake was to appear in many of Bell's future novels.

The second earlier experience on which Bell drew in *Green Fire* was contact during his year in New York (if only by reputation) with the inventor M. I. Pupin. Although an electrical engineer, Pupin was close to the mathematicians at Columbia and was in fact a charter member of the American Mathematical Society. The notation "Jevic = Pupin" appears on the flyleaf of Bell's personal copy of a later edition of *Green Fire*. Clearly he was familiar with the life of the inventor—a gifted Serb who had arrived in America with nothing but five cents in his pocket and a red fez on his head. Unlike Bell's fictional Jevic, Pupin received recognition and reward from his adopted country and was able to use his influence and money to benefit society. Bell's Jevic, also a Serb and "a scientific intelligence of the first order," is obsessed with punishing the world for its derision of him and the red shirt he was wearing when he arrived in America. "Once and forever," he vows, he will make the world "respect intelligence."

Green Fire is often dismissed by science fiction critics as John Taine's "variation on the mad scientist theme," but it is much more than that, as Everett F. Bleiler points out in *Science-Fiction: The Early Years*.

"While much of the book suffers from poor exposition, Jevic's account of his early life, which takes up about forty pages, is fascinating. Taine sets out to give a rationale to the mad scientist of fiction, and to some extent he succeeds, despite defects in the novel."

I do not know how long Bell spent on this first attempt at a novel, but in his Bell profile Hughes wrote, "so fertile is his creative impulse, so swift his powers of composition, that three weeks suffice him for the production of a full-length novel." This statement is supported by all the holograph copies of Bell's novels on which there are dates at the beginning and the end.

Green Fire is important as Bell's first attempt in a new field, but the amount of space I have devoted to it, even without giving any idea of the plot, is out of all proportion to the amount of time that he devoted to its writing. I cannot possibly give a comparably detailed description of the mathematics he was doing. It is simply too technical for the story of his life that I am writing. But I want to emphasize that only *three weeks* in Bell's year were devoted to his recreational writing; the other *forty-nine weeks* were spent on the teaching of mathematics and mathematical research. He was in no sense a college professor who spent his evenings writing for amusement and profit. Coincidentally, for the year in which he wrote *Green Fire*—there is a listing by Bell himself of what he was doing mathematically.

"Since April 15, 1919, I have carried out and completed the following specific investigations in mathematical research," he reported in April 1920. He then proceeded to list sixteen topics, all but one of which later appeared as published papers. "In addition, I have carried on four more [investigations], which are not yet complete and, therefore, not listed." He also mentioned "two long papers still with the referees who have not yet (after 12 months) reported on them." This time lag illustrates the discrepancy that often exists between the date of a piece of research and the date of its publication.

Three of Bell's sixteen investigations lay outside the theory of numbers. "The Gedankenwelt of Dedekind," never published, was "a short note attempting to show that Dedekind's famous proof of the infinity of the world of ideas is fallacious." "On proofs of mathematical induction" was a note "pointing out fallacious reasoning in the method as summarized by Poincaré." "Parametric solutions for a fundamental equation in the general theory of relativity and gravitation, with a note on similar equations in dynamics" was in mathematical physics.

The year in which Bell made his report was a year in which the general theory of relativity had captured the imagination of the public as had no other scientific theory since Darwin's. Earlier in 1919 it had been tested experimentally; and the results, confirming Einstein, had been made public on November 6. The name Einstein shortly became as familiar to the general public as the names of the "Big Four" who had recently hammered out the peace treaty at Versailles. It was commonly stated in the press that only ten—or was it twelve?—men in the world understood Einstein's revolutionary theory. "Understanding" became a public obsession, and in this connection Bell was to play a role. I learn about it only when I come upon the following addendum to Martin

Gardner's column in *Scientific American* on what are now known as "Bell numbers":

> Alan Watton, Sr. in a letter to *Scientific American* (July, 1978) disclosed that Eric Temple Bell's first published article on mathematics was in *Scientific American* in 1916. The paper won honorable mention in a contest for the best explanation of relativity theory written for persons of average intelligence.

Like so many pieces of information about Bell, this one is wrong. There is no such paper in *Scientific American* in 1916, and no such contest; yet when I look in the magazine for July 1978, I find a letter from an Alan Watton Sr. worded exactly as cited!

The date 1916, so soon after the publication of his theory, does seem a little premature for a contest to explain it. So I proceed to search through subsequent issues of the magazine. At length, in July 1920, I find what I am looking for.

An American millionaire named Eugene Higgins was offering a prize of $5000 "for the best essay on the Einstein theories so written that it may be read by an ordinarily intelligent person with no special mathematical training." *Five thousand dollars* was a great deal of money—Bell's salary as an assistant professor at Washington was $2000 a year. But there was also the challenge: to explain in 3000 words or less "as simply, lucidly, and nontechnically as possible" the new theories of relativity and gravitation. Not a great deal of time was available. In the next four months three hundred people from all over the world, many of them very well known, submitted essays. Excited reports appeared in each issue of *Scientific American* as to the progress of the contest. The deadline was November 1. The prize—there was one, the entire $5000—went in the end to a Mr. L. Bolton of the British Patent Office.

"The interest taken in the [Einstein] contest from the first, alike on the part of prospective contestants and of those who expected only to read the essays which it would bring forth, has been all that we could have anticipated," the Editor of *Scientific American* reported. Unfortunately, so he shortly had to admit, clear and concise as Mr. Bolton's essay was as a statement of the postulates of relativity theory, "it was somewhat lacking in explanation." As a result the magazine, which had planned to print it and some of the other essays in its columns, decided that instead it would publish a book on the contest. It was hoped that with the inclusion of a preface—and bits and pieces from various essays as well as some other essays printed in full—all in addition to the prize essay itself—the volume would enable a general reader to come to an understanding of Einstein's theories. Bell's essay was one of fifteen chosen to be printed in their entirety in *Einstein's Theories of Relativity and Gravitation,* a volume advertised as "[giving] the clearest thoughts of the greatest authorities—'Einstein at a glance.'"

Bell's essay, which was apparently his first attempt at explaining a scientific subject to a general reader, was titled "The Principle of General Relativity: How Einstein, to a Degree Never Before Equalled, Isolates the External Reality from the Observer's Contribution." It is well organized and well written. His examples are taken from everyday life. But the Bell personality is completely suppressed. I doubt that anyone reading the essay would guess that it was written by the man who later wrote *Men of Mathematics.*

It was probably the same summer as that of the contest, but at least in the same year, that Bell permitted himself another three-week "fictive orgy" (Hughes's description). The result was *The Purple Sapphire.* Unlike *Green Fire,* which is set in New York and New Jersey, *Sapphire* takes place against the kind of exotic background that was to be a distinguishing characteristic of many John Taine novels and develops Soddy's idea of an advanced civilization that makes a fatal mistake.

The most extensive analysis of John Taine's work is that by James L. Campbell Sr. in Bleiler's *Science Fiction Writers.* Campbell, a professor of Victorian literature, describes *The Purple Sapphire* as "exhibiting a marked technical virtuosity in combining a number of disparate Victorian story formulas," and concludes that Bell is "skilled enough as a romancer to make this composite structure succeed."

In style and execution the novel conjures up for most readers H. Rider Haggard. In fact, Taine Bell remembers his father's advising him at a later date that if he wished to learn to write he should take as a model a writer he admired and then try to write like him. Haggard is familiar to Taine as one of his father's examples.

Bell had written *Green Fire* in 1919 and *The Purple Sapphire* in 1920. In 1921 he would write still another novel. All three of these, each in its own way, would play an important role in the world of "John Taine." But in the much larger world of E. T. Bell a more important event was the publication at the beginning of 1921 of "Arithmetical Paraphrases." It appeared—in two parts—in the *Transactions of the American Mathematical Society,* which was a journal that the Society reserved at that time for long and fundamental papers.

"Here we have something both novel and profound," Durst tells me, giving the paper three stars, his highest rating. "The genuinely new idea is the notion of *paraphrase,* which Bell sets out in Part I with his proof that the trigonometric functions may be replaced by arbitrary functions with the proper 'parity,' i.e., evenness, oddness, symmetry."

The work on paraphrases had grown directly out of Bell's study of Jacobi's arithmetical (as opposed to analytical) application of elliptic functions and the later development of the subject in the hands of others, especially Liouville. This approach was *discrete* as opposed to *continuous* and hence appealing to Bell. As mentioned earlier, Liouville had suggested in the title to his memoirs that his formulas "peuvent être utile" in the theory of numbers. And useful indeed they could be! Dickson, the leading American number theorist, had devoted a whole chapter in the second volume of his history of number theory to "this remarkable collection of formulas" and had showed, as an example, how easily an important theorem of Jacobi's could be derived from Liouville's so-called first formula. Now Bell, using his method of paraphrase, was able to derive the first formula with much greater ease from formulas of the theory of elliptic functions, particularly the Jacobi theta function.

"It's like pulling rabbits out of hats," Durst marvels, pointing out under what he calls "Bell's slick tricks" that he actually derived the first formula under his "simple illustrations."

"This part of the argument is quite in the spirit of Euler, who was the supreme master of formal manipulation of infinite series and products," Durst explains. "Bell's method is very systematic and general; he is able to derive all of the Liouville formulas, and a great many due to others (including a host of so-called class number relations or recursions, created first by Kronecker) as well as an inexhaustible supply of similar formulas of his own."

Dickson himself later enthusiastically summarized Bell's idea of paraphrase, "roughly" as follows:

"Bell obtains from the [theory] of elliptic or theta functions many identities in [trigonometric] functions. From *any* such identity, he deduces, by a formal process (proved valid once for all), an identity involving a nearly *arbitrary* function (which may be an even or odd function or possessing certain parities). This function may be taken to be one of many special types, each choice yielding a *theorem* in *the theory of numbers.* The theorems of the last type so far deduced by Bell are new for the most part—& in the rare cases when not entirely new are

L. E. DICKSON sketched while giving an address at Strasbourg in 1916

such that earlier proofs are either lacking or by tricks. Hence Bell obtains a vast array of theorems in the theory of numbers from analysis as source, & hence introduces a most powerful tool, easily applied (& without tricks)."

It was Dickson's opinion that the value of the work for the theory of numbers could "not be overstated."

For Bell the year 1921 had begun very well, and it was to continue to be a very good year. In January and April "Arithmetical Paraphrases" appeared in the *Transactions*. In May he received a raise of $200, which brought him to $2800 a year, and a promotion to associate professor. In July, while also teaching in the summer session, he wrote a novel dealing with the nature of time. He called it *Remembered Worlds*. It would not appear for another ten years, and then as *The Time Stream*.*

The Time Stream shows the influence of the passage from Soddy that Bell quoted at such length in *Green Fire*. Much of it derives from the myths and legends cited by Soddy, particularly the legend of the Fall of Man and the myth of the Undying Fire. But it also reflects in fictional form the paradox of the philosophical meaning of "now" and the geometrical ideas on space and time that Hermann Minkowski had presented at Cologne a few months before his premature death in 1909: "Henceforth space by itself, and time by itself, are doomed to fade away into mere shadows, and only a kind of union of the two will preserve an independent reality."

In addition to autobiographical details about Bell's days in San Francisco, the novel also contains material from his childhood in San Jose. The son of the heroine is, unknown to her, a child from the planet Eos. She is concerned that he writes poetry, has strangely vivid dreams, and makes up games for a collection of lizards—his favorites being the horned toads.

Bell felt that the idea of this novel was a new one, as he later explained to John Macrae, the president of the E. P. Dutton and Company: "the possibility of human beings having cognizance of their own immediately continued existence in time, past, present, future, and the ability at times to see the whole span of history complete. [In the novel] I trace the story of the golden age, not in this world or even in this system of stars, but in another which has perished, but which has left its trace on the present and which still 'exists' in time."

The story begins in San Francisco during the fateful April of 1906 and involves a group of friends and acquaintances who are suddenly thrown into the stream of time. Drifting through time, which is circular and symbolized by the snake with its tail in its mouth, they return to a planet so ancient that the very atoms composing it have disintegrated. They then become observers and participants in the destruction of the succeeding civilization of the planet Eos. They return to San Francisco and experience the earthquake of April 18, 1906, much as Bell experienced it, and in the same part of the City.

Everyone who has read *The Time Stream* agrees that it is a confusing story. It lacks the classical unities of *Green Fire* and the narrative

---

* After a reference to the original title I shall refer to the John Taine novels by the title under which they appeared in print.

skill of *The Purple Sapphire*; yet, as is almost invariably noted, "there is a fascination," it is "oddly compelling," "haunting, poetic, unforgettable" with "a startling emotional depth and power." Bell thought then, and continued to think, that it was the best thing he had written "in prose."

I mention to Mike Ashley, the science fiction bibliographer and author of a recently published biography of Hugo Gernsback, that *The Time Stream*, acknowledged as one of the earliest stories in which the nature of time was treated, was written long before it was published.

"What year would it have been, then? About 1922?" he queries by return mail. "This information would be particularly pertinent because *The Time Stream* is regarded as one of the important early works of sf, but its publication date of 1931 ... puts its development within the context of evolving science fiction which had emerged as a separate genre in 1926 when Hugo Gernsback launched *Amazing Stories*. If John Taine had been writing about his concept of time travel as early as 1922, then it makes his work even more revolutionary because he presaged some of the ideas postulated (independently) by writers in the pulps in the 1926–31 period, rather than following on from them as is currently felt."

Bell did not try to sell *The Time Stream* since he felt, as he later wrote to Macrae, "... it would take a rather peculiar public, I believe, to get the drift."

In September 1921 he began his tenth year of teaching at Washington. Along with his usual courses that fall he was to teach three quarters on the Theory of Relativity. The University was beginning to attract more academically inclined students, and Bell's course was numbered 207-208-209, indicating that it was meant for graduate students. Several departments, although not yet mathematics, would be awarding Ph.D.'s in June.

By 1922 Bell seems to have been reasonably satisfied with his situation at Washington. A happy hermit, as his friend Hughes described him, working away at what he wanted to do at the table in his living room with its golden walls. An "adoring wife" (this is Hughes's description). A certifiably bright little boy (according to the psychologist who had been consulted on the subject when Taine was three). Yet he would have been unusual if sometimes he had not felt that with a bibliography of nearly thirty mathematical research papers and a growing reputation in the theory of numbers—mark the enthusiasm of Dickson!—some other university should have wanted him, too.

JESSIE "Toby" BELL with her son, Taine

Recognition

MEMO   PRESIDENT'S OFFICE   May 11, 1922

Professor E. T. Bell reported that he had been offered a full professorship at a salary of $4000.

President Suzzallo offered to increase his salary to $3500 and to promote him to full professor at the beginning of the next academic year.

The offer reported above was Bell's first from another university in the ten years since his Ph.D. After "some consideration" he informed President Suzzallo that Washington was "much to be preferred" to whatever institution it was that had made the offer. (It is never named in the records.) He then proceeded to give his reasons. First, "the congenial surroundings" made it possible "both to teach profitably and, through the standards sets [of journals] in the library, to carry on research in mathematics." Second, the University was beginning to establish a place for itself "on the mathematical map" of the United States, "and I should like to be here when it finally arrives." Third, there were in the mathematics department "five brilliant students whom I would not willingly leave for a mere five hundred dollars." Fourth, there seemed to be a "new tone" at the University since Suzzallo's arrival: "It is now a place where serious intellectual and scientific work not only is made possible, but is actually done."

At the beginning of that summer he took another three-week respite from mathematics and wrote another novel. A holograph copy of *Desmond*, today listed at $5000 in the catalogue of a dealer in science fiction books and manuscripts, is inscribed "Written May 28–June 19,

1922." It is quite different from *The Time Stream*. The latter is science fiction, putting forth the idea that people have simultaneously existing lives in time. *Desmond* is a mainstream novel that debunks the idea that there can be communication between the living and the dead. Suffering from amnesia after a railway wreck, Desmond is believed killed. His father, a distinguished scientist who is duped into believing he is receiving messages from him, writes a book about the experience. The novel closely paralleled the case of the physicist Sir Joseph Lodge, who had written a book about his communications with his son, who had died in France.*

Bell's book ends with Desmond and his friend Bicknell making plans to found a company to publish scientific books in attractive and readable form at nominal cost. As Bicknell tells Desmond, "Such a thing as your father's *Glimpses of Immortality* would not be possible in a world that had a high school education in ordinary science." In later years, when Bell was well known as a scientific popularizer, he told people that he started writing fiction in the hope that by showing publishers he could write salable material that would sell he could interest them in his serious writing. If there was such a motive at the time, it was that he was probably hoping to interest publishers in his long poetical works. It is true, however, that very early he was emphasizing the need for popular treatments of science.

His "fictive orgies" did not affect the quantity or quality of Bell's mathematical research. The titles under "Bell, E. T." in the international review journal continued numerous. As classes began in September 1922 there was "Periodicities in the theory of partitions" in the *Annals* and "Anharmonic polynomial generalization of the numbers of Bernoulli and Euler" in the *Transactions*. Both were in the third area of Bell's work that Durst classifies under the heading Special Numerical Sequences and Functions.

But at this time Bell's most important mathematical investigation, recently completed, was his paper on "Euler Algebra." In this fundamental work, to which Durst gives three stars, he laid out in detail the foundations of the program he had introduced in his 1915 paper in the University of Washington series, and described the methods he had employed in the earlier work in a much more formal way. The paper

---

* The novel was retitled *Damon* and later *Called Back*, as Bell fiddled with dates and names, trying to make it publishable but never attacking the problem of the similarity to the Lodge story.

contains the fundamental general treatment of the algebraic properties of generating functions that he used to explain the development of his theory of numerical functions, especially factorable or multiplicative functions. He also proved that analytical questions (in particular, convergence of the infinite series involved) were entirely irrelevant for the tasks at hand. In short, as Durst emphasizes, he generalized the theory of generators to arithmetic functions. This last is perhaps the best example of Bell's eccentricity as a mathematician who was at home only in the discrete and who could dismiss the analysts' continuum as a *chaotic smudge*.

"Analysts often tend to see generating functions as sources of information about the growth of numerical functions as the 'variable' n increases, so they require that proper attention be paid to the questions of convergence that are required to justify manipulation of the series," Durst explains. "There is a fascinating passage in Hardy & Wright (*An Introduction to Number Theory*) where they explain—in small print— that one *could*, if one really wished to do so, develop what they are doing, with their high-powered analytical techniques, in a purely formal way, but they seem to suggest that it might be a waste of effort, since it would be just a 'translation into a different language ... and that ... we gain nothing by the translation.' They certainly do seem to treat the whole thing like a strange idea altogether."

With the development of the high-speed electronic computer, however, this "eccentric" paper of Bell's, along with half a dozen other papers of his, was to become an indispensable part of the baggage of every modern combinatorialist.

"The paper has an independent interest as well," Durst points out. "Here we find Bell at the threshold of what came to be called 'modern' or 'abstract' mathematics in the nineteen twenties and thirties. At the time he wrote this, the terminology had not settled down to the standard collection of words we use now, but his treatment is very much in the abstract or modern manner. He just doesn't have the modern terms at his disposal yet. In short, he was ready before they were."

It happens that for the same year in which "Euler Algebra" appeared, I am able to obtain a recollection of Bell by one of his students. In response to a query in the University of Washington alumni publication, I receive a letter from Dr. George R. Nagamatsu, now eighty-seven years old but presently combining his degrees in electrical engineering and medicine for a third career as president of the Society for Urology and Engineering. Dr. Nagamatsu writes that he came to Seattle in 1922, "the hick son of a Japanese immigrant." In Snoqualmie he had been the salutatorian at a four-teacher school with twenty-seven students, and he found himself unprepared for the "brutal anonymity" of large classes, sophisticated classmates, and distant professors.

It was his good luck, so he writes me, that Professor Bell had volunteered "to lower himself" to teach the needed additional class in differential calculus to which he had been assigned.

"Like a child observing his father," he observed the professor: "the crisp formal and yet informal manner in which he spoke, his aquiline nose and other facial features, the crop-type cane that he carried, the cut of his clothes, the cap—all told me he was an admirer of Sherlock Holmes."

Bell scribbled the equations for the day, perhaps five, on the blackboard. The students assumed ten minutes would be allotted to each equation, but Bell considered two minutes enough. The remaining eight minutes he devoted to such questions as *What is space?* Some of Nagamatsu's classmates expressed the wish that they had been assigned to another teacher, but Nagamatsu was enthralled. He spent hours in the library trying to answer the questions Bell posed. Then, to his dismay, he realized that the time spent in the library should have been spent on the calculus. Panicky, he asked for a private meeting.

"His pipe came out as soon as class ended. The whole left-hand upper drawer of the desk was his ash tray: he would empty his pipe, tap it at the top of the drawer, scrape it of its ashes, then tap it again to make sure it was empty—all this while he talked to me. When I told him of my confusion and lack of background, he began to explain in a ruminating sort of way, recalling facts so elementary that he had had no occasion for years to reach for them, putting them together for my benefit."

Nagamatsu barely passed the course ("the only *D* I ever got"), but he writes that he has always been grateful for what Bell taught him in those short and irregular sessions and has tried to apply it in his own teaching as a professor of surgery: "That a really great man is a kindly person, that his first mission as an educator is to reach down to the student, letting him know that he himself was in that position at one time, and to stimulate him to go on and not to despair."

By this time the calibre of students in mathematics was much higher, as Bell had told Suzzallo. During the past year a young woman had written a paper for her master's degree "of such exceptional merit that it was accepted at once for publication by one of the international scientific journals."* Students who had gone to eastern universities had made remarkable records.

Of the five exceptionally brilliant students, the one that Bell found most exceptional was a husky young man from Montesano, Washington, who had the additional recommendation of being of Scottish descent. Young Howard Percy Robertson was so excited by the big university that he greeted almost every new experience or piece of knowledge with the exclamation "Hot Diggety Dog!" Bell, who loved American slang, immediately picked up the expression and for the rest of his life called Robertson "Hot Dog" or "Diggety."

Robertson's father, who had died when the boy was fourteen, had been a civil engineer; and when in the fall of 1919 the sixteen-year-old Robertson had arrived at the University, he had planned to become an engineer like his father. To his surprise—all this comes from autobiographical notes—he found that the only part of engineering he really liked was the underlying mathematics. His mother, however, had also expected him to become a bridge builder like his father, and she did not look with approval on his decision to transfer from the School of Engineering to the School of Science. According to Robertson's daughter, Marietta Fay, it was Bell who convinced the mother of the wisdom of the change by telling her that if her son became a mathematician, he would earn $10,000 a year. Many years later he was to confess, "At the time I didn't believe it myself."

In Robertson's last year as an undergraduate his interest in relativity was stimulated by Bell's course on that subject. When, not yet twenty, the young man graduated, Bell suggested that he remain at Washington for another year and then go to the young California Institute of Technology for his Ph.D.

The only person I am able to contact who knew Bell during his Washington years is Dr. Nagamatsu, but I am able to talk to children of people who were Bell's colleagues then. They recall that their parents and others on campus talked with disapproval of the couple's unconventional dress, their language, their jokes, the nicknames they gave to others and often would not reveal, their unwillingness to mix. Most of all they disapproved of the way they were raising their little boy, the fact that Taine called his father and mother "Romps" and "Toby,"

---

* Echo D. Pepper's master's thesis was published in the *Tôhoku Mathematical Journal*. She went on to take her Ph.D. at Chicago and to teach at the University of Illinois, Champaign-Urbana.

that he was being "pushed" by his parents, who seemed determined that he must be a genius. (In 1922, at the age of five, Taine had entered the third grade, having been privately tutored since he was three.) But particularly shocking was the fact that the Bells did not send Taine to Sunday school. In fact, they apparently gave him no religious education at all. The story was often repeated of how the little boy passing a church and seeing a cross on the steeple asked, "Why is de plus up dere?"*

(I can give no reason for Bell's often extravagantly expressed hostility toward various religions, and religion in general. His only personal statement on the subject was that his father had forced him as a child to memorize long passages from the Bible. Throughout his life, however, he enjoyed coming up in unexpected situations, including the mathematical, with an always appropriate quotation from scripture.)

The unfriendly campus comments cited above should not to be construed to mean that the Bells did not have friends on campus. Taine recalls a number of these for me. The philosopher C. J. Ducasse (after whom Bell named a character in *The Time Stream*) had come to Washington the same year as Bell. He and his wife, Mabel, who was an artist, were the Bells' first friends on campus. I have already mentioned Glenn Hughes, who was by this time a member of the faculty. The statistician Harold Hotelling spent a year or so at Washington and named his son after Bell. Irving Gavett, a colleague in the mathematics department, owned an island cabin where the Bells on occasion joined him and his wife for a weekend. There was also Louis Peter de Vries, a Dutchman in the Foreign Language Department, and his wife, Mary, who was an especially good friend of Bell's. "He always attracted good-looking, interesting women," Taine says. The de Vrieses had a boat that slept four, and with them the Bells enjoyed overnight trips on the water. Then there was "Puck," or Margaret Petit, a ballet dancer. Later, married to a French chemist, she became the mother of Leslie Caron.

One can imagine the cluckings on campus if in the spring of 1923 the news got around that Professor Bell had sold a *novel*. I don't know whether it did, or whether the identity of "John Taine," like that of "J. T.," was still being protected. At any rate, without a stipulation to that effect, Bell signed the contract with E. P. Dutton and Company

---

* This story is authenticated in Toby's handwriting in a section of her address book labeled "Sayings of Taine."

MARGARET "Puck" PETIT, later Mme. Claude Caron

for *The Purple Sapphire* on March 3, 1923, a month before he presented "Euler Algebra" to the San Francisco Section. He was understandably excited. Within two weeks he was writing to Dutton, "I am getting a bit anxious about the production of my novel. . . . I do not like to trouble you, but when do you think it will be out?"

The acceptance stimulated Bell to set aside more time for recreational writing; however, according to Hughes, "lest it absorb too much time and energy [he] held it in check by a rigid routine." Only twice a year did he "permit John Taine to control his mind and his pen."

By the time Bell signed the contract for *Sapphire* he may have already completed *In Search of Life* (ultimately published as *The Greatest Adventure*). It appears to have been written between semesters in 1922–1923, since the hero, Dr. Eric Lane, is—like Bell—in his fortieth year (and very satisfied with his life).* *Adventure* was the first of the John Taine novels to have a biological theme. These are considered "his best and most interesting work," according to Brian Stableford, author of the "Taine" entry in the *Science Fiction Encyclopedia*, edited by Peter Nicholls.

As *Green Fire* is essentially a book about intelligence, *The Greatest Adventure* is a rumination on scientists and science. (Bell later used the title for a commencement address he gave at Caltech in 1928.) The story begins with a creature—"neither bird, reptile or fish, [but] an incredible mongrel of all three," which could not have been dead more than fifteen minutes when tossed up in an eruption of subterranean oil at the South Pole along with a number of pictograph-inscribed rocks. The independently wealthy Dr. Lane (Bell's scientists often have "patents" that support them in leisure) promptly sets off for the Pole accompanied by his adventurous daughter and her young man.

There, through the pictographs, they learn the history of a civilization that discovered the secret of creating life but turned out such monstrosities that it committed mass suicide in order to destroy its creations. But they were not able to destroy all of them . . .

In the summer of 1923 (June 17 to July 14) Bell indulged in another "fictive orgy." The result was the never published *Satan's Daughter*, the story of an evil pagan princess ruling an island isolated since pirate days and rich in gems and uranium.

---

* This dating for the novel is further supported by the fact that Roald Amundsen, then preparing to leave on his expedition to the North Pole, was a guest at the University's Research Society in May 1922, and Bell refers to his expedition in the book.

With the publication of "Euler Algebra" and the sale of his first novel, 1923 had been an eventful year for Bell; 1924 was to be even more eventful. There was an offer from another university (again not named), apparently $5000, to which Suzzallo responded with $4500. Again Bell decided in favor of Washington. Then Dickson, impressed by Bell's work on paraphrases, invited him to spend the summer at Chicago, which was unusual in having then, as now, a lively summer session.

Bell greatly admired Dickson, whom he liked to call "the Texas cowboy." There is a little of the look of Texas in portraits of Dickson; but in fact he was born in Iowa, the son of a banker, and merely took his bachelor's degree at the University of Texas while working as a chemist for the Texas Geological Survey. Dickson had devoted much of the past nine years to a three-volume *History of the Theory of Numbers*. In this work his purpose had been, not to trace and explain the sequence of development, but simply to state what had been accomplished in the theory. It was an approach that appealed to Bell. As he later wrote in *The Development of Mathematics*, "From his own experience and that of others ... the professional mathematician suspects that often what looks like an anticipation after the advance was made was not even aimed in the right direction."

Bell watched Dickson work during the summer of 1924 and several subsequent summers always with amazement. The older man finished his mathematical work in the morning and spent the afternoon and evening playing bridge. He never seemed to look at a book or journal, "and wrote *Modern Algebraic Theories* out of his own head," so Bell later marveled to H. S. Vandiver.

Dickson's enthusiasm for "Arithmetical Paraphrases" was to play a determining role in Bell's receiving the most significant honor of his professional career. Two years earlier, in 1922, the American Mathematical Society had announced a prize in memory of Maxime Bôcher. It was believed to be "the first mathematical prize in our country to be given at regular intervals for research in pure mathematics." Because of Bôcher's editorial connection with the *Transactions*, the award of $100 was to be made for a memoir published in that journal during the past five years by a mathematician under forty. In 1923 George D. Birkhoff of Harvard University, who was by then considered the leading American mathematician, had received the first Bôcher Prize. He and Dickson and H. S. White had then been designated to choose the recipient of the next Prize, to be awarded in 1924. Initially the three

GEORGE D. BIRKHOFF
won first Bôcher Prize

committee members inclined toward a memoir by Solomon Lefschetz of Princeton. Before making their final decision, however, they decided to ask the opinion of an authority in Lefschetz's field, the Italian mathematician Francesco Severi. After reading Severi's report on the work of Lefschetz, Dickson wrote doubtfully to Birkhoff.

"I got the *impression* that Severi regards the memoir [of Lefschetz] as in *part* a further development of *ideas originated* by Poincaré &c, but not suffic. developed by Poincaré. Certainly a glance thro' the memoir shows that it depends very much on earlier work. I don't mean to detract from its high value. But I wish only to raise [the] question if we should not also seriously consider the *highly original* long paper by E. T. Bell on Arithmetical Paraphrases in the same volume . . . ."

Dickson had heard Bell present his work "very clearly" during the Chicago summer session just past. He was "fully convinced both of the high order of merit & [the] originality of [Bell's] theory, which has almost unlimited applications to the theory of numbers." He felt, as did also White and the colleagues at Chicago, that it was "highly worthy."

"I know little first hand of Lefschetz' memoir; but from Severi's evaluation I am willing to vote it a tie with Bell." Lefschetz was still under forty, but this would be the last chance for Bell, who had been thirty-eight when his memoir was published but was now forty-one.

"And I cannot easily bring myself to vote against his most remarkable & original fertile work."

So the matter stood in the summer of 1924.

That same summer Bell's mother died. He had not seen her in twenty-two years. From his memorabilia there is little indication of contact with her. On occasion, in later years, he told friends that his family in England did not know where he was but that he "kept track of them." At any rate, after Helen Bell died, dividing what she had equally among her three children, he would have had to have contact with his siblings, at least through solicitors. In addition, with the termination of Helen Bell's life estate in Penybryn, the family home of James Bell Sr. passed to the three children of James Jr. The house may have been leased for a while, but it was then sold.

On September 1, 1924, as Bell began his thirteenth year at Seattle, his mother was buried, not in Norwood with her husband, but in Bedford. It would be interesting to know if Bell ever learned from his sister of the proud words on his mother's gravestone that would ultimately reveal the secret of his boyhood in the Santa Clara Valley: *Widow of James Bell of San Jose, California*. But then—were they proud words? Or were they even Helen's words? I wonder. The ultimate responsibility for the placement of the inscription over her mother's grave would have had to be assumed by Enid, who always described her father as "cruel to his wife and children." James Bell Jr. had never become an American citizen. He had spent the greater part of his life in the British Isles. His wife had returned his body to England for burial—at great cost to herself and her children. It seems strange that in these circumstances she would have identified him almost thirty years after his death as being "of San Jose, California." Yet if Lyle had not learned of the inscription on his great-grandmother's gravestone, it is doubtful if I or anyone else would ever have learned about Bell's boyhood in America.

In the month following his mother's death Bell's novel *The Purple Sapphire* appeared. He inscribed the first copy "To Dog Toby/With the respects of the Author/John Taine = E. T. Bell." Reviews reported in the *Book Review Digest* weren't generally "raves." "Only Sir Rider Haggard, that we can remember, has so piled Ossa upon Pelion as regards adventure and lurid description" (*Boston Transcript*). "We can readily forgive the story for being improbable; but what we cannot forgive is the author's elephantine attempts at humor" (*Outlook*). "The author is better when writing about scenery than the mysteries of love" (*Literary Re-*

view). But the *New York World* found it "a book of action, brushed with color and tinted with love," and the *Christian Science Monitor* called it "an excellent diversion."

Almost immediately—Bell stated that it was in the fall of 1924 (which is an odd time for him to have been indulging in "a fictive orgy")—he wrote *The Iron Star*, his first "devolution" novel. The story begins when a doctor notes "apish" characteristics in a new patient, a former missionary. Sure that these are connected with the man's African life, the doctor and the father and son physicists, Big Tom and Little Tom Blake (nicknames Bell had for himself and his son), set out for that continent. There they unravel the mystery of a band of apelike creatures who have "devolved" from men. The leader of the band, perhaps the most memorable character in the John Taine œuvre, is a figure of tragic stature—Campbell McKay. Once a noted Scottish explorer, he has not succumbed as completely as his followers to the addictive effects of "the iron star" and still remembers what it was like to be a man.

In addition to the publication of a first novel by John Taine, the year 1924 had seen the appearance of a dozen research papers by E. T. Bell, an invitation to lecture at one of the most distinguished universities in the country, and participation in the International Mathematical Congress as the official representative of the University of Washing-

SOLOMON LEFSCHETZ
shared Bôcher Prize

ton. The climax of this eventful year came appropriately at its end. In December, at the annual meeting of the American Mathematical Society, Birkhoff announced the joint recipients of the Maxime Bôcher Memorial Prize:

E. T. Bell of the University of Washington

and

Solomon Lefschetz of Princeton University

BOB AND ANGELA ROBERTSON when he was a graduate student at Caltech

Choices

Bell always felt, according to Taine, that his former Stanford professor H. F. Blichfeldt was responsible for the fact that he remained at Washington as long as he did.

"How did your father think Blichfeldt kept him at Washington?"

"My father thought that Blichfeldt passed around negative things about him."

"What kind of things?"

"That he was coarse and uncouth."

"Not a gentleman?"

"That's right. But then he began to get out in the mathematical world and people saw that he was not like that, he was acceptable."

After the announcement of the Bôcher Prize—when his work on paraphrases had been described by Birkhoff as "a fundamental contribution to number theory"—the word "acceptable" was something of an understatement in regard to Bell. But it was a full week *before* the announcement of the Prize that Robert A. Millikan, the Chairman of the Executive Council at the California Institute of Technology and the winner of the Nobel Prize for Physics the previous year, dispatched letters to the three men he had been informed were the leading mathematicians in the country: George Birkhoff, L. E. Dickson, and Oswald Veblen. His purpose was to inquire as to their opinion of E. T. Bell as well as of a member of his own faculty. The letter to Veblen, a copy of which is in the Library of Congress, is probably almost identical to what he wrote to Dickson and Birkhoff.

OSWALD VEBLEN
consulted by Millikan

"I am trying to get as correct a judgment as I can about the standing among mathematicians of this country and the world of two men
whom I think you know,—the one is our own Dr. Bateman and the
other is Professor E. Bell of the University of Washington," Millikan
wrote to Veblen. "From the standpoint of physics and mathematical
physics we are fairly competent here at the Institute to form judgments
in which we have some confidence, but from the standpoint of mathematics I feel keenly my own incompetence."

If Bell and Bateman were members of the National Academy of
Sciences, or were being considered for membership, then Millikan
thought he would have some idea of their status among their fellows.
Since to his knowledge neither man was an American citizen* and
therefore ineligible, he would like to ask: if they *were* citizens *would
they be likely to be nominated* to the Academy?

There are a couple of things of interest about Millikan's query.
Bateman, an English mathematician, had been on the staff at Caltech
since 1917, when it was still the Throop College of Technology.
Millikan had been Chairman of the Executive Committee at Caltech
since 1921. It seems odd that he had no idea of what he already possessed in Harry Bateman, a mathematician of quite extraordinary abil-

---

* Bell had in fact been naturalized in 1920.

ities. But Bateman was always to be undervalued at Caltech. As for the query about Bell, according to members of the Bell and the Robertson families, Millikan's interest in him had been aroused by the enthusiasm of a graduate student at Caltech—young "Bob" Robertson, Bell's "Hot Diggety Dog." It appears that in admitting his inability to judge the calibre of mathematicians Millikan was not being falsely modest.

All three of the men he contacted responded promptly. Both Bateman and Bell were among those who had been proposed for the next year's ballot of the Mathematical Section of the National Academy, so Millikan was informed by all three. Veblen wrote that he did not know anything about Bell's work, but he was aware that Dickson had a very high opinion of it "and Dickson is, of course, a leading authority in that field." Birkhoff "[could] vouch for the power displayed" in the paper for which Bell had received the Bôcher Prize, but as editor of the *Transactions* he had seen papers by Bell that were outside the theory of numbers (he was probably referring to the foray into relatively theory) "and I have come to the conclusion that when he departs from his specialty, his work is not always of high order." It was, of course, Dickson who gave Bell the most glowing recommendation. He was "an A-1 mathematician of very exceptional ability in research of high order on fundamental subjects." If Millikan was thinking of giving Bell a call, "I will say you could hardly, I think, get a better man. ... But," he warned, "we shall at first opportunity try to get him at Chicago."

Nothing followed immediately upon these responses, but the following summer of 1925—when Bell was again at Chicago—that university offered him a permanent full professorship at $6000, a third again as much as he was getting at Washington. Chicago vied with Harvard and Princeton as an outstanding center in mathematics. Bell himself considered it the most outstanding. The three leaders of American mathematics to whom Millikan had written had received all or most of their training at Chicago. It was a most flattering offer.

When President Suzzallo was informed, he commented to Toby, who had remained in Seattle, that he supposed finally Bell would be unable to resist Chicago but inasmuch as he seemed to like it in Seattle and seemed to want to stay in that climate he, Suzzallo, would go as high as $5000.

In spite of the fact that Bell considered the climate in Chicago "vile for at least 500 miles in every direction," he was finding his sum-

The Archives, California Institute of Technology

HARRY BATEMAN: "a vivid, at times almost romantic, imagination"

mer there "very pleasant in every way," he wrote to a Washington colleague. He had two classes of advanced students, some of them very
good, "but none with the independence of our best Washington students." When he was informed by Toby of Suzzallo's counteroffer, he
wrote to the President that, once again, he had decided to remain at
Washington.

"Of course, I have always the feeling that they are not going to
stop at $6000," Suzzallo responded a little wearily, "but at least I have
expressed my goodwill and my personal appreciation of you by going
to the limit of my capacity here."

Bell was quite happy with his new salary at Washington, but he
was not in Seattle during much of 1925. Although he complained that
all this "running around the country" interfered with work, he continued to turn out mathematical research. There were to be a dozen
papers published in the following year.

Novels by John Taine also continued to accumulate. In the summer of 1925, while Bell was still at Chicago, Macrae wrote that Dutton
was interested in a book to follow *The Purple Sapphire,* which had done
well and was shortly to be published in England. Early that June (so
stated on the holograph copy) Bell had written the novel titled *Red and
Yellow.* It began with a wealthy American woman on her way to Turkey
to adopt two Armenian orphans, and ended with an eruption "of foreign pests well known in Russia and Asia" sweeping over "exclusive
America." It was a timely story, but Bell did not mention it to Macrae.
Instead he suggested *The Iron Star* or *The Greatest Adventure.* He tried
not to repeat himself, he told Macrae, "still keeping to the main idea of
giving everything a scientific setting." Both stories were "exciting."

"The whole darned lot of us are on the upgrade now," he exulted
to Glenn Hughes. "If occasionally we strike a rock in the road it is
nothing but a temporary jolt to keep us humble."

Bell spent the fall semester of 1925–1926 as a visiting professor at
Harvard. He found the colleagues there quite different from those at
Chicago.

"[T]he majority (on a first impression) are more intrigued by
settling a question in a gentlemanly fashion rather than by wringing
the brute's neck and having done with it. To put it another way: the
Chicago crowd is in the game to win at all costs; Harvard would rather
pose as a coterie of distinguished amateurs and save its savoir faire."

But the library was "a dream" and Boston, with its crooked streets,
"a joy." The countryside reminded him of England, and from Harvard

Bridge the city looked something like London. "All in all it is the best city I have struck in America."

Bell returned to Seattle to spend Christmas with his family. On the fourth day of the new year he took up his father's schoolboy copy of Julius Caesar's *Commentaries on the Gallic War* and printed, very lightly and in pencil, at the bottom of the title page:

Jan. 4, 1926

7:58 p.m.

In mem. ETB

30 yrs.

After 1926 he no longer memorialized his father's death in the Caesar. But he continued to be sensitive to its passing—in 1929 noting in the margin of the novel he was writing: "Jan. 4–33."

Having declined an invitation from Rice Institute to spend a semester or two there, Bell returned to the Washington campus. After his semester in the East he was armed with information about the salaries being paid on that coast, and he immediately became involved in "a glorious fight . . . over the rotten salaries paid to our younger men."

Unknown to Bell, but at almost the same time, President Suzzallo received a letter from President Nicholas Murray Butler of Columbia University sounding him out on the subject of E. T. Bell. Those on Butler's faculty who had known Bell from his days as a graduate student thought highly of him, but others would like to hear Suzzallo's views as to scholarship, productivity, and research that would "fit or unfit him" as a colleague.

Suzzallo responded straightforwardly. Bell was one of the outstanding scholars at Washington. No one on the faculty outranked him. Although not much interested in elementary teaching, he was very interested in teaching "the beginnings of mathematics" to bright students with the potential for becoming scholars.

"Like most gifted men, he holds his own views rather tenaciously and might be said at times, like other real geniuses, to be somewhat temperamental, but I believe that he would be a helpful and cooperative colleague in a group of advanced scholars. He would return cooperation for cooperation [although] he is so devoted to research that he is sometimes a little impatient in having his time and energy used up by activities that keep him from what he wants to do."

Suzzallo realized that he was not going to be able to keep Bell for long, "and I would wish him to go to so distinguished a place as Columbia University if he would care to go."

After receiving Suzzallo's letter, Butler wrote directly to Bell of the possibility of a newly created research fellowship at Columbia with a salary of $7500 a year. It was a dream of a position, as described:

"The purpose is to add to the staff of the University a productive scholar in the field of Mathematics whose chief interest is in advanced work, in research and in publication and who will be able and willing to devote his entire time to his own researches and to publication of their results, as well as to the guidance of such advanced students as come to him in the hope or expectation of going forward to the degree of Doctor of Philosophy with Mathematics as their subject of major interest."

Almost simultaneously with Butler's letter came a letter from Robert Millikan saying that if Bell would be interested in joining the staff at Caltech, he (Millikan) would try to be in Seattle in a couple of months so that they could talk the matter over. Caltech had something to offer that Columbia, like Chicago, did not, namely, *climate* along with scientific prestige. Bell realized that if he wanted a position at Caltech, he could not wait for Millikan to come through with a counteroffer.

He responded immediately and directly.

"Any possibility of joining the staff at the Institute would appeal to me most strongly. I do not wish to seem forward in this matter; but I should hate to neglect any opportunity which might make it possible for me to join you, and the force of present circumstances makes it necessary for me to put all my cards on the table . . . ." He then described to Millikan the state of the negotiations with Columbia and the salary being suggested. "Now of course I realize that allowance must be made in salary for the lower cost of living on the Pacific Coast compared with the Atlantic, and working conditions at the Institute would also be different. However, if it is possible to come to some agreement which is mutually satisfactory to the Institute and to me I would much prefer the Institute to New York for several reasons, the most important being that at the Institute there is without doubt the most notable body of men in the mathematical and physical sciences in America, and further the program of the Institute is by common consent unique in its promise."

Millikan realized that he could not wait. He responded immediately that he would very much like Bell to join the staff at Caltech, but he stated no conditions and made no definite offer.

"We are in the very deuce of a pickle," Bell complained to R. G. D. Richardson of Brown University. "Both places professionally are first class. Pasadena for natural surroundings is our idea of heaven. What are we to do?"

Richardson urged Columbia.

"You would wield great influence there. ... I might say in confidence that Birkhoff was offered a position there, but Harvard met the situation by giving him the research professorship. ... I think it is not unflattering to you that you should be second choice after Birkhoff."

Richardson's letter came too late to have any influence on Bell's decision. Once again he had chosen the West over the East, accepting $6000 at Caltech rather than $7500 at Columbia.

"The deciding factors were climate and living conditions," he explained, but he added, "To my way of thinking Pasadena has the 'most interesting' possibilities of any place in the U.S."

The years Bell had spent at Washington had been remarkably productive for him as a mathematician, although he was always to complain that Moritz, the "Pterodactyl," purposely arranged his teaching schedule so that he would not have uninterrupted time for research. Moritz's version was somewhat different. He told his nephew, the mathematician H. A. Arnold, that people at Washington made fun of Bell as he crossed the campus with his cap and his pipe, his cane and his red smoking jacket. "But I told them, 'That man is a first-rate mathematician, and we want to keep him here, so I always arrange his classes so he can have time for his research.'"

I have not come across a single complimentary remark by Bell in connection with Moritz. He frequently wrote, however, that the one thing at Washington that enabled him "to go on" was "the really excellent mathematical library"—which could only have been the result of efforts by Moritz as department chairman.

In his fourteen years at Washington, after a slow start in the first five of those years, Bell had published eighty research papers. His sponsors in the National Academy of Sciences, proposing him for membership in its Mathematics Section, circulated a list of his "more important papers" among the fifteen members of the Section.* Seventy-three of Bell's eighty papers were included on this list as "more important."

Durst classifies Bell's mathematical work at Washington under

---

* G. D. Birkhoff, H. F. Blichfeldt, G. A. Bliss, A. B. Coble, L. E. Dickson, L. P. Eisenhart, Edward Kasner, Solomon Lefschetz, G. A. Miller, E. H. Moore, W. F. Osgood, W. E. Story, E. B. Van Vleck, Oswald Veblen, and H. S. White.

three different headings. These are Arithmetical Functions and Theories; Forms, Elliptic Functions, Paraphrases; and Special Numerical Sequences and Functions. The third classification is unusual in relation to the other two in that the distinguishing theme is, not a general theory from which a number of results follow, but rather *a tool*. This tool, the umbral calculus, was one that Bell had first met two full decades earlier as "the symbolic method" in Lucas's *Théorie des Nombres*. The somewhat poetic name currently used—the umbral calculus—came from J. J. Sylvester, the English mathematician after whom Bell named a principal character in *The Time Stream* (a character the details of whose life bore certain resemblances to Bell's own).

In the course of his career Bell wrote three expository articles on the umbral calculus. The first of these appeared during 1925–1926 in the *American Mathematical Monthly* under the title "An algebra with singular zero." Although in this paper Bell emphasized the underlying algebraic properties of the system rather than its applications, he used the umbral calculus frequently as one of his machines (so Durst refers to them) to crank out a seemingly infinite supply of mathematical relations between a great variety of special sequences and functions.

In the spring of his last year at Washington, asked again by the administration to turn in a report on his work, Bell listed as the research on which he was currently engaged: a general theory of numbers, a general theory of functions important in mathematical physics, the arithmetic of logic and relations in general, and "Algebraic Arithmetic" (capitalized and in quotes).

"Algebraic Arithmetic" was the subject and title that Bell had selected for the Colloquium Lectures he had been invited to deliver the following summer at the meeting of the American Mathematical Society. These traditional lectures had their origin in lectures that the great German mathematician Felix Klein had delivered at Northwestern University following the World's Columbian Exposition in Chicago in 1893. According to the American Mathematical Society, which later adopted the idea, the purpose of its Colloquium Lectures was to present "before discriminating and interested auditors the results of research in special fields" and to promote "personal acquaintance and mutual helpfulness ... among the members in attendance." The lectures were then published in book form so that even mathematicians not present could derive benefit from them.

It was an honor to be invited to deliver the then week-long series of Colloquium Lectures, and Bell had already begun to work on what he was going to say. In a paper published before he left Washington he had begun to bring together a variety of ideas that interested him and that related to the greater part of his mathematical achievement to date. In this paper, which was titled "Irregular Fields," he managed to touch base on most of the topics he would take up in his Colloquium Lectures.

During Bell's fourteen years at Washington, in addition to an impressive amount of mathematical work, he had done a great deal of recreational writing. He had published two books of poetry and one novel. He had also written nine other novels. The last two—*The Gold Tooth* and *Quayle's Invention*—were apparently written between the semester at Harvard and the beginning of the first semester at Caltech. I do not have exact dates, but *The Gold Tooth,* which begins in the Boston Public Library, probably preceded *Quayle's Invention,* in which the heroine brings her father to Pasadena for his health. Both novels deal with the theme of the Philosopher's Stone, the alchemists' dream, but otherwise they are quite different. The first is in the style of *The Purple Sapphire*—exotic background, lost civilization, great natural catastrophe—while the second is in the straightforward style of *Green Fire.*

Bell's years at Washington also culminated in a significant event for mathematics at that university. At graduation in June 1926 the University of Washington awarded its first Ph.D. in mathematics. The recipient was Clyde Myron Cramlet, who was to be for many years a member of its mathematical faculty. The title of Cramlet's thesis was "Invariant tensors and their application to the study of determinants and allied tensor functions." Given Bell's interest in relativity theory, I am only a little surprised to find that it was he who had signed off the thesis.

At this time, preparing to leave Seattle for Pasadena, Bell was as concerned about Washington and the mathematical situation there after he left as he was about the future of mathematics at Caltech. The same mathematician figured in both these concerns.

At Chicago the previous summer Bell had met a National Research Council Fellow who had impressed him greatly. This was Aristotle Demetrius Michal, a young man of Greek descent. Born in Smyrna in Asia Minor, Michal had come to the United States when he was twelve years old. At Rice Institute he had been a student of Griffith Evans, who had passed on his own interest in functionals to his student. Michal was very bright and very ambitious. He also had a charming and beautiful wife, Luddye, whom he had married the day he received his Ph.D. and she, her A.B.

There are in the Caltech archives nine letters from Bell to Michal, several of which were written during the spring and early summer of 1926 while Bell was still at Washington. In 1926 Michal was twenty-seven years old. His NRC Fellowship was about to expire, and Bell encouraged him to think about Washington.

"Our university is, in my opinion, slated to be the best on the Coast (outside of Pasadena in its specialties) inside of 10 years. A man under 30, with persistence, tact, and the indurated, obstinate will to work, could not do better than come here. An Easterner would criticize us unmercifully, and most likely would be unsympathetic. But a man who has had experience in the newer parts of the States would grasp our point of view and make good."

He did not want to lead Michal to think Washington was better than it actually was.

"Your [mathematical] stimulus must come entirely from yourself. But, even at that, you will find a well organized group of men in all lines (about 30, our 'Research Society'), powerful in University administration, whose main interest is research, and who do everything in their power (which is considerable) to encourage the younger members of the staff to produce."

The teaching load would be twelve hours.

"This is *not* excessive," he assured Michal. The work would be elementary except for one graduate course, "and it is very well compacted in the schedule so as to conserve a man's time and energy. If a productive man insisted, he could cut this load to either 10 or 8 or even 6 hours, *provided he had proven that he really was turning out high class research.* Suzzallo told me I need teach only 6 hours if I liked."

In spite of Bell's case for Washington, Michal decided in favor of Ohio. Disappointed but undaunted, Bell nevertheless continued to make plans for the young man.

"My foothold [at Pasadena] is not yet firm. But Millikan told me last week that he hoped to build up the *pure* math. I hope to get some really live young men into the dept. I shall first try for Robertson (Millikan wants him), and then for you."

This—as he prepared to leave Seattle for Pasadena—was Bell's plan for the future of mathematics "at Tech."

FOUR

the prime years

All This . . .

On September 28, 1926, the *California Tech* carried the headline: COL-LEGE STAFF INCREASED BY TWO NOTED MEN. Bell was described as having "an enviable reputation as a teacher" as well as being "a distin-guished research man in mathematics." John Taine was not mentioned. The other addition to the faculty was Chester Stock, a paleontologist, "interested primarily in mammalian fossil remains in tertiary forma-tions west of the Rockies."

At that time Pasadena—the home of the California Institute of Technology—was a sheltered little town backing against the bulk of the San Gabriel Mountains. Even the Southern Pacific did not stop there. One had to get off at Alhambra or Glendale and take the street-car the rest of the way. Wealthy midwesterners—most of them from Iowa—came to winter or to retire. Not much happened after the sun went down unless there was an amateur production at the Community Playhouse. The big events of the year were the Tournament of Roses on New Year's Day and the Iowa Picnic. But the exotic flowers, shrubs, and trees might have come out of one of the lost paradises of which Bell wrote. Flaming bougainvillea crept up walls, jacaranda stood in blue pools of fallen blooms, the cup of gold could be heard opening as one watched. Highways were bordered by oleander in three colors. (There would not be a freeway for fifteen years.) Especially fascinating to Bell was the century plant, which was reputed to bloom only once in a hundred years.

The Bells bought a modest two-bedroom frame bungalow next to a large vacant lot at the corner of South Michigan and San Pasqual. Today that intersection, part of the campus of Caltech, is in the middle of the mall that leads to Beckman Auditorium, but in the fall of 1926 Bell strolled across the street to the campus. On the way to his classes, which were held in the building that had housed the Throop College of Technology, he passed "Gates," the famous chemistry laboratory over which Arthur Amos Noyes presided. Noyes had been the first "big name" that George Ellery Hale, the Director of the Mt. Wilson Observatory, had lured to Throop. The next had been Robert A. Millikan, "The Chief" to everyone on campus except Bell and his wife, who privately referred to him as "Pa."*

Administration at the Institute was more flexible than at the University of Washington. There was no president—Millikan's title was simply Chairman of the Executive Council. There were also no departments. "The Chief" was a gifted man who could have been a success in many different fields. According to Jesse W. M. DuMond, one of his later students, he had initially inclined toward a career in physical education. Fortunately he had developed a passion for scientific knowledge instead. He had come to Caltech under a contract that explicitly provided that he was not to be required to take its presidency, that he was not to be expected to spend time raising funds, and that his first and most important job was to build the best department of physics of which he was capable. He never took the title of President, and the Norman Bridge Laboratory of Physics was to become world famous; but the second provision of the contract had already gone "out the window" (in his words). In 1924, recognizing the need for the Institute to move forward with the rapidly growing area that surrounded it, he had come up with the idea of the Institute Associates—a hundred men "who would be both able and eager to put in a thousand dollars apiece each year for a period of ten years in order to push this enterprise along before it is too late."

The Institute's catalogue for the year of Bell's arrival listed nineteen full professors in addition to Bell and Stock. Emphasis was primarily on physics and chemistry, as the number of National Research Council Fellows in those two subjects testified. The situation of mathematics was modest, although not unduly so in comparison with the Institute as a whole, which was still very small. In addition to Bateman

---

* Millikan, the physicist; Hale, the astronomer; and Noyes, the chemist, were later immortalized in a large joint portrait that Bell, with his propensity for nicknames, promptly christened "Tinker, Thinker, and Stinker"—a designation still used by the irreverent at Caltech.

R. A. MILLIKAN as Nobel Laureate for Physics in 1923

and Bell, the mathematical faculty consisted of one full professor (H. C. Van Buskirk—"Van Buzzard" to Bell), one associate professor, one assistant, and one instructor. These handled the elementary mathematics teaching. As at the University of Washington there were heavy demands for mathematics instruction at Caltech but little for mathematical research. There was not a single NRC Fellow in mathematics on campus when Bell arrived, and only one teaching fellow. This last was Morgan Ward, who had graduated in 1924 from Berkeley, where he had been inspired to go into mathematics by the number theorist D. N. Lehmer. Later, visiting his mother in Pasadena, he had been shown around the Institute by Bateman, who had urged him to come south for graduate work. Ward was to be Bell's first Ph.D. at Caltech, and Caltech's first Ph.D. in mathematics.

The first year Bell taught Infinite Series and Mathematical Analysis and conducted a seminar on Algebra and the Theory of Numbers. He wrote to Hughes: "The work here is light, except that immoral suasion compels me to attend more lectures than are decent [he was always to dislike learning from lectures]"—and to Michal: "They let you alone, and blush when they ask you to teach 4 hours + 1 course in research."

His principal associate in the graduate work in mathematics was Harry Bateman, who was listed in the catalogue as Professor of Mathematics, Theoretical Physics, and Aeronautics. Bell was to find Bateman (as he wrote after the latter's death in 1946) "a singularly modest and gentle man." He was "always ready to place his skill and his knowledge at the disposal of others, with no thought of personal credit," possessing in his many contributions to mathematical physics "a vivid, at times almost romantic, imagination." Although both men had emigrated from Great Britain and were less than a year apart in age, they had had quite different careers. Bateman had won an entrance scholarship to Trinity College, Cambridge. A series of impressive achievements at Trinity had followed, then study abroad in Göttingen and Paris. In 1910 (while Bell was teaching mathematics at Yreka High School) Bateman had emigrated to the United States and a position at Johns Hopkins University. At the same time Bell was taking his Ph.D. at Columbia, Bateman took a Ph.D. at Hopkins although, according to the biographical memoir of the Royal Society, he had already published some sixty papers, among them his "celebrated researches" on the Maxwell equations and his "monumental report" on integral equations. In 1917, unappreciated at Hopkins in terms of salary and professional status and unhappy with the collegiate atmosphere of the University, he accepted an appointment at the Throop College of Technology. Lack of appreciation apparently continued, for it was this man of whose reputation in the mathematical community Millikan had felt called upon to inquire when he was also inquiring as to Bell's.

There was complete agreement between Bateman and Bell as to what was needed in mathematics at Caltech. The first job was "to convert Millikan to pure mathematics"—these were Bell's words in another letter to Michal—"he thinks he is already converted, but he hasn't guessed the half of his salvation." At the same time the mathematical library had to be built up. In this respect the Institute was the inferior of Washington, for it did not subscribe even to such an indispensable journal as *Crelle's*.

One other colleague should be mentioned, although he was not a professor of mathematics. This was Paul Epstein, Professor of Theoretical Physics, a brilliant man, Russian born, who had taken his Ph.D. in Munich with Arnold Sommerfeld, had taught in Moscow and Zurich, and had come to the Institute in 1921 at the invitation of Millikan, one of the latter's first appointments.

Beginning a career in Pasadena, Bell was determined to show his mathematical mettle. At the meeting of the San Francisco Section he presented a paper, "Arithmetic of logic" (the long delayed Part II of "An arithmetical theory of certain numerical functions"). Durst, who gives "Arithmetic of Logic" one star, describes it as a "fascinating paper [in which] we see another of Bell's attempts to generalize the fundamental theorem of arithmetic, in this case by constructing its analogues in Boolean algebra." The paper is one of several that Durst feels, "at least in hindsight," look very much like precursors of lattice theory, which was not to be created until some half dozen years later.

Office of Public Relations, Caltech

PAUL SOPHUS EPSTEIN
an early Millikan choice

I have hoped that by the time I get to the years that Bell spent at Caltech I shall be able to contact at least a few people who can give me a sense of the man as a person—in short, that I will find Bell in Pasadena. From the Caltech catalogue I obtain the names of several, now in their eighties and nineties, who were NRC Fellows or students in 1926–1927. Unfortunately most of them recall no very specific contact with Bell at that time. An exception is Linus Pauling, who was a research fellow in chemistry the year that Bell came and a colleague for many years thereafter. Just before the great celebration of his ninetieth birthday, Pauling writes me that he remembers very well how Bell tried to encourage him to become a mathematician, suggesting that he work on Fibonacci numbers, but that he was not interested in mathematics except as it applied to chemistry and physics. His most vivid memory is of a commencement address Bell gave.

"He started his address by saying that the statements he made should be attributed to him alone, and not considered to be opinions of Caltech or other people at Caltech. I listened eagerly to hear what iconoclastic statements he would then make. I was disappointed that he made none, and I went away feeling that he must have been insecure and unsure of himself to have begun his address with an apology."

During his first years at Caltech, Bell continued to turn out an impressive number of papers. There were to be eleven published in 1927 and twelve in 1928; yet he complained that all his spare time was being devoted to "those blasted Colloq. lectures." He wanted to include a great deal in them on the subject that he called "Algebraic Arithmetic" and that he felt was *his subject*. He laid it out in his introduction:

"Intermediate between the modern analytic theory of numbers and classic arithmetic as developed by the school of Gauss, is an extensive region of the theory of numbers where the methods of algebra and analysis are freely used to yield relations between integers expressed wholly in finite terms and without reference, in the final propositions, to the operations or concepts of limiting processes. This part of the theory of numbers we shall call *algebraic arithmetic*."

The boundaries of the region were not sharply defined, "nor is it desirable that they should be, as power comes from flexibility." After all, in his opinion, almost any part of arithmetic could be profitably employed in any other. But there was a large, growing, and somewhat uncoordinated body of results, with many aims and methods peculiar to itself, which fell into neither the classic theory of numbers nor into the modern developments of the analytic and algebraic theories. For this reason he felt that Algebraic Arithmetic "opened up" many suggestive opportunities "for systematic exploration."

The manuscript for the lectures continued to grow. In them he wanted to bring together in a unified whole all three areas of mathematics in which he had been working during his years at Washington, but he also wanted "to outline a few promising directions in which progress [might] be made toward classifying, extending and generalizing methods and results." He had to cut relentlessly: "Three times I have been mistaken in the estimated number of pages allowable."

Forced to concentrate to such an extent on his Colloquium Lectures, he did not indulge, it seems, in his usual fictive orgy in the summer of 1927. However, another John Taine novel, *Quayle's Invention*, was published that year by Dutton.

Bell's contracts with Dutton and other publishers were almost the only personal papers he saved. The date on the contract for *Quayle's Invention* is November 1926, not a surprise in view of the publication date. The surprise is the fact that the contract contains as well an agreement on the part of Dutton "to publish within the next three years, and not more than two in any one year, four other works of the same nature and already written." The four works were *The Iron Star, Green Fire, The Gold Tooth,* and *The Greatest Adventure.* It seems clear that the publisher must have had quite a success with *The Purple Sapphire.*

The plots of almost all John Taine novels are both complex and complicated, and any attempt on my part to summarize them would result in a disproportionate amount of space being devoted to an activity to which Bell himself devoted comparatively little time. The detailed summaries by Bleiler in *Science Fiction: The Early Years* are recommended for the interested reader who does not have access to the six novels published by Dutton. Bleiler himself almost despairs of summarizing *The Gold Tooth,* explaining to his readers, "The plot is so complex that it is best approached by the leading characters." These involve four groups, two of which are Japanese (military modernists and conservative traditionalists) and two American (represented officially by the Secretary of State and independently by a brash young archeologist). All this may sound like a reasonably well organized scientific adventure, but when one adds four romances, two of them interracial, a good and beautiful pagan princess, ancient scrolls that appear to embody the wisdom of *De Re Metallica,* a bit of satire in the style of Sinclair Lewis, a hunt for dinosaur eggs a la Roy Chapman Andrews, and a dog that resembles Dickie Doyle's "Dog Toby" on the covers of *Punch,* one can see Bleiler's problem.

The Gold Tooth is interesting for anyone trying to understand the racist aspect of much of Bell's science fiction that offends modern readers and has caused many to turn away from his novels. The heroine, Geraldine Shortridge, leaves her middle-class home to join the man she thinks she loves, a Japanese aristocrat named Satoru Okada. Bell is obviously impressed by his handsome, intelligent, aristocratic—and tall—Japanese, a type he may have met for the first time at Harvard and one quite different from the "Jap" laborers he knew in San Jose and San Francisco. (There was, however, a club of some thirty Japanese students at Stanford when he was there.) But the novel is not a true interracial romance. Okada is taking advantage of Geraldine to hide a precious "gold" tooth, with which she (at his instruction) has had one of her own teeth replaced. When Okada arranges that she follow him to Japan on a Japanese vessel, she comes in contact with a large number of his countrymen. She is repulsed by their strangeness and difference, in short their "Japaneseness," and cured of her infatuation.

"There is something in race prejudice after all," Bell concludes. "Otherwise we should never hear of it. It may be wrong and stupid, but it exists. That is the fact."

In *Quayle's Invention,* on the other hand, Bell idealizes the aborigine who befriends Quayle as morally superior to the "civilized" people of whom he writes. The banker Cutts, the father of Sheila, whom Quayle loves, has left that young inventor to die on the shores of "Westralia." Cutts's excuse is that Quayle is "a menace to civilization" in whose hands chemistry has become "an engine of destruction waiting only [his] will to release it and abolish . . . the slow progress of . . . centuries." But Cutts himself soon joins with the brilliant and unscrupulous Barton to use Quayle's invention to rescue himself from financial disaster. Although Quayle originally sees the monetary value of his invention as a way of proving himself worthy of Sheila, "The probable consequences of [it] had not greatly troubled him. To create, to invent, to put the new science to new use, that, and that alone sufficed him."

In his critique of John Taine's work, James L. Campbell Sr. sees three main actions as giving Quayle its particular meaning. These are Cutts and Barton's conspiracy; the campaign to foil it by Quayle, Sheila, and Dr. Burdette (a Pasadena psychiatrist treating the now guilt-ridden Cutts); and Quayle's growth from self-destructive revenge to moral redemption. Campbell finds "this last component, though superficially upstaged by the exciting action of the two adventure plots, . . . the most important, since it unifies the three plots into a single artistic whole. Taine's account of Quayle's progress from psychological fragmentation to moral wholeness is detailed, well-defined, and central to the novel's controlling morality, which argues that only through love and social responsibility can mankind find peace and happiness."

It is hard to tell how much of this admirable thematic organization is due to Bell and how much to the editors at Dutton who had a hand in revising *Quayle's Invention* for publication—a revision that Bell described to John Macrae, the president of the company, as "a mistake."

Dutton trumpeted *Quayle's Invention* as "a story as brilliant, as daring, and as wildly speculative as any Jules Verne ever wrote." *The Book Review Digest* reported Will Cuppy as writing in the *Herald-Tribune*: "This story is not skimped in plot or in any particular which could make it better in any way." There is no review of *The Gold Tooth* mentioned in the *Digest*.

During the years that saw the publication of these two novels, Bell continued to hold to his plan for the development of a strong mathematics program at the Institute.

"I hope you do not think I have forgotten what we talked about so many times," he wrote to young Michal at Ohio. "Our situation here is as follows. Millikan thinks that he believes in pure mathematics, but he doesn't. He hasn't the least conception of what it is all about. Hence when I ask—or hint—that we need more men, & say we shall have to pay respectably, he simply does not get the point. He actually thinks— it sounds incredible—that one man, namely myself—can be competent to handle both the algebraic and the analytic side of mathematics."

Young Morgan Ward, "a very brilliant man," had been added to the staff as a part-time instructor "at a salary that should make a decent man blush." Bell's former student, H. P. Robertson, having received his Ph.D. at Caltech, was in Europe on an NRC Fellowship; but Bell still had plans to get him back to Caltech.

"I plan for him to do the great work of connecting the math & theoretical physics. In the latter, with Bateman & Epstein we are already as good as (if not better than) anything in the country. But before we can do anything in mathematics we must have a man who is expert in *modern analysis* . . . ." For that, someone like Michal would be needed. "I have done all I know how in an unofficial way to urge that we acquire men under 30 years of age who have real promise. M. will listen— a little—to this, as he knows from actual examples that I have never failed yet to pick comers before they arrive. I have done my darndest to make him get you, but he hesitates . . . ."

It was the old problem. Millikan wanted men who were already notorious—"that is exactly the proper word."

Finally, in the summer of 1927, Millikan came through with an offer of an assistant professorship for Robertson. Although that young man was currently studying in Germany, he was as eager as ever to rejoin his old teacher and friend; and he promptly accepted Millikan's offer, which would have him assume his new position in September 1928. Thus by the end of his first year, Bell had brought to fruition half his scheme for mathematics "at Tech." He had Robertson "on hold." He still had to get Michal.

Bell again spent the summer of 1927 at Chicago. A reaction to his lectures there comes to me, thirdhand as it were, from an interview with the number theorist Ivan Niven conducted by Albers and Alexanderson in 1981. At that time Niven recalled a conversation with Walter Gage, who had been a student at Chicago in the summer of 1927, his instructors there having been W. C. Graustein and Bell. The lectures of the former were exemplary in precision and organization, but Bell's lectures were "something else again." Proving a theorem at the blackboard and finding himself stuck, he would print in large letters DE-TOUR and then draw a thick arrow circling back to some earlier and more satisfactory point from which he could proceed with the proof.

Niven asked which method Gage had found preferable.

"Both," Gage replied. "Two Grausteins or two Bells would have been too much. One of each was just right."

Following the Chicago summer session, Bell went to Madison to deliver his Colloquium Lectures on "Algebraic Arithmetic." This was a meeting of the American Mathematical Society more noted today as the first at which a woman, Anna Pell-Wheeler of Bryn Mawr, was a Colloquium Lecturer. Both Wheeler's and Bell's lectures were extremely successful—more than five hundred persons attending.

Dickson hailed *Algebraic Arithmetic,* the book in which Bell's lectures were ultimately published, as being of "marked originality [and] of vital interest to advanced students in various branches of mathematics." He found it "a systematic attempt to find a unified theory for each of various classes of related important problems in the theory of numbers, including its interrelations with algebra and analysis." It was abstract, but only because it was then possible for Bell "[to give] the essence of a vast array of concrete results traced to their true sources and easily deduced from these few sources."

He concluded his review with the firm statement: "This original and scholarly book is an honor to American mathematics."

Although Dickson especially emphasized Bell's use of the paraphrase method, many of Bell's ideas in the lectures went back to the first paper he had published after he came to the University of Washington—the paper that had been so unduly neglected because it had appeared in the local series. Apostol tells me that he much prefers that paper, "An arithmetical theory of certain numerical functions." *Algebraic Arithmetic* is simply not easy to read "in spite of the fact that Bell is supposed to be a good expositor."

"Do you think that is because (as he complained to Michal) he felt that he was so limited as to space?"

"That may be. But the principal results are formulated in such a general form that the key ideas are obscured. Perhaps for this reason the book has not been widely read, and many of its contents have been rediscovered independently in later years by other mathematicians."

GIAN-CARLO ROTA learned late of Bell's work

A mathematician not so easily turned away by Bell's book is Gian-Carlo Rota of the Massachusetts Institute of Technology, who feels that some of Bell's insights in it were very much ahead of their time. Rota's focus, unlike that of Dickson, is on those parts of interest to combinatorialists rather than to number theorists. In this regard *Algebraic Arithmetic,* so he writes me, "has been for many years the book of seven seals, and I hope to write an exposition of what it contains, now that I believe I begin to understand it, after twenty-odd years of perusal."

I ask Rota if he feels that the difficulty of the book comes from the fact that Bell may have approached mathematics with some of the exuberant imagination that he let loose in his novels.

"I think that is quite possible," he replies. "It is another way of talking. There is a certain 'other worldliness' about his writing. In fact, his presentation may be responsible for the fact that his ideas, so many of which are very original and very deep, have not had the circulation they deserve."

I comment that it would be ironic if Bell, who was famous for explaining mathematics to laymen, would have failed to communicate his own mathematical ideas to other mathematicians.

"Yes," Rota agrees. "Yes. It would be ironic. *Algebraic Arithmetic* is a very unusual mathematical book, but privately I doubt if anyone has read it all the way through."

"Have you?"

"No. Not yet."

For a variety of reasons Bell's work has often not been known to people who should know it. About the same time that *Algebraic Arithmetic* appeared, the *Journal of the Indian Mathematical Society* contained a paper by a Dr. R. Vaidyanathaswamy in which he put forth as a new result a discovery that Bell had published a dozen years before. His theorem, Vaidyanathaswamy wrote, was "a fundamental one in the whole theory of Arithmetic Functions" and it was "rather surprising" that it did not seem to have been utilized before, "[since] the existence of the inverse of any multiplicative function throws a great deal of light on many known theorems on Arithmetic Functions."

Bell responded tactfully but specifically to the author.

"Your recent note . . . was of particular interest to me, as I have used the same method for several years in precisely your form of it, also in an extended sense. Thinking that you might like to hear of some of this work, especially in view of your remarks concerning your main result . . . I send you a short description of my own approach [and also] a list of some [in fact, there were fourteen] of the papers I have published on this and on closely allied topics . . . ."

Bell's reply appeared in the *Journal of the Indian Mathematical Society,* 17 (1928), 249–260, under the title "Outline of a theory of arithmetical functions in their algebraic aspects."

"That article ought to be required reading for anybody about to publish some of these things for the nth time," says Durst with a certain amount of exasperation—"if not in the interest of conserving paper, at least in order to assign credit for things that are known (some from long ago) and claim appropriate credit for whatever is novel."

A few years later in a review, Bell also expressed himself on the subject of the duplication of results: "There may be some merit in establishing known theorems all on one's own, but there is more in consigning the results to the wastebasket . . . ."

In June 1928 Bell returned to Chicago—but for the last time.

"The Institute . . . doesn't like it very well," he explained to Glenn Hughes, "& has given me two very handsome raises with the plain hint that I lay off in summer."

It is clear, two years after his arrival there, that Pasadena had turned out to be every bit the paradise Bell had expected it to be. He was producing a record number of mathematical papers. His novels were beginning to appear regularly. He was making very close to that ten thousand dollars a year he had so rashly promised Mrs. Robertson that her son would make.

Yet even in paradise—

In a letter to Professor Smith acknowledging the receipt of Smith's critical revision of *Le Comput Manuel de Magister Anianus,* Bell mentioned that, reading the collected works of Ramanujan in preparation for writing a review, he couldn't help noting a striking kind of abstract similarity between Anianus and the Indian mathematician who had so recently died at thirty-two.

"Both were half-pioneers; neither, a thousand years from now, will be more than a historical relic."

Their works brought home to him "the utter futility of all mathematical endeavor." It was not, of course, a subject which he expected to concern a historian of mathematics like Professor Smith. And yet—"it does concern the creative mathematician."

A. D. MICHAL about the time Bell met him at Chicago in summer 1924

Dr. Bell and Mr. Taine

In the fall of 1928 Herbert Hoover was running for president, promising "a chicken in every pot and a car in every garage," and Bell was thoroughly happy with his professional situation at the California Institute of Technology.

"Conditions for research here are better than any I have known anywhere in the U.S.," he wrote Michal, "the man who keeps alive is well treated. . . . Our policy is to get one outstanding, experienced man in each department (such as T. H. Morgan in biology) to start things going, and then to hang on to our own most brilliant products, & bring in young men from the outside."

Robertson was already listed in the catalogue as "Assistant Professor of Mathematics on Leave," and although "Pa" did not yet see the necessity for adding another man to the mathematics staff—an analyst in the person of young Michal—Bell was sure that within a year he could win him over. For the present Bell himself was teaching Modern Analysis. The physicist H. Victor Neher, who took the course from him, got the impression that he knew the subject pretty well "but was an easy grader."

Bell's extra-professional life in Pasadena was also very satisfactory. He had a love of growing things, perhaps inherited, and he enjoyed gardening in Southern California. Already (with the permission of the owners) he had taken over the big vacant lot next door. His garden, however, was not what most people envisioned as a garden. He was not interested in beds and borders, so the result was more like a jungle.

He was also making a determined effort to learn to paint with watercolors. In this endeavor his teacher was Roger Hayward, an original man of varied talents who had designed a number of civic buildings in the Los Angeles area and was well known as a sculptor, watercolorist, and scientific illustrator. Although Hayward had no degree in astronomy, he loved the subject, wrote several papers on it, and later (with Caspar Gruenfeld) completed the largest model of the moon ever built.

Although Bell still followed his rule of saving half his income, with "two substantial raises" he had money to satisfy his fondness for Chinese bowls and vases and first editions of Victorian poets, which he searched out with the assistance of his friend Jacob Zeitlin, a well-known Los Angeles book dealer. The volume he desired most but never found, according to Zeitlin, was Francis Thompson's *Poems* containing "The Hound of Heaven." He also fancied the work of John Davidson, a Scot, known in his time as "the Poet of Anarchy."

Young D. H. Lehmer, the son of the University of California mathematician D. N. Lehmer, came down from Berkeley during his junior year to talk to Bell about a problem of mutual interest—the application of elliptic functions to number theory. He found Bell friendly and

D. H. "Dick" LEHMER
as an undergraduate*

---

* Pictured with the first of his many special-purpose computing machines, known as sieves, this one constructed with nineteen bicycle chains of various lengths.

The Archives, California Institute of Technology

THOMAS HUNT MORGAN
his lab inspired *Seeds of Life*

accessible but Pasadena a sleepy little town "where they rolled up the sidewalks when the sun went down." He thought that Bell was happy in such a situation.

There were in Pasadena, as there had been in Seattle, a select group of friends: Edwin Hubble, the astronomer; Graham Laing, the economist; Thomas Hunt Morgan, the biologist, a neighbor; Chester Stock, the paleontologist; Richard Chase Tolman, the physical chemist, another neighbor. Bell was always attracted to men in other fields, particularly to what one might call "active scientists." Although he agreed completely with Gauss that mathematics was the queen of the sciences and never considered another career, unless possibly a literary one, there is not a mathematician in any one of his novels.

Of his Pasadena friends the closest were Edwin Hubble, who was at that time providing the first observational evidence for the expansion of the universe, and Hubble's wife, Grace. She was to become one of his closest women friends, to be thanked frequently (often with her husband) for her assistance in many of his later books on science. When I learn that her personal journals are in the Hubble papers at the Huntington Library, I hope that I can obtain from them a firsthand portrait of Bell in his early Caltech days, and an analysis of his character. But I am disappointed. She was, as her husband described her, more excited by ideas than by people. Her references to "Romps" are almost invariably to things that they talked about. From her journals, however, I do get an idea of what it was about her that attracted Bell. Again it came in her husband's words, quoted in her journal: "Edwin said that a woman to be attractive to him must have a certain spiritual quality, she did not have to know about many things, but she must have a certain clarity of mind."

In spite of the number of these friends it was Bell's desire, in young Lehmer's view, "simply to be let alone to do his work." And there was a lot of work. In terms of number of papers published, the first years at Caltech saw the peak of his productivity. None, however, of the work in multiplicative diophantine analysis, which was to be his fourth main area of interest, according to Durst's classification, had yet appeared.

He continued to be concerned about the future of mathematics at Caltech. He felt that young Robertson, who was still in Germany, had not been producing the kind or the amount of work that he had expected of him. If Millikan demanded "notoriety," Bell demanded "productivity." In his letters to Michal—by this time endeavoring to persuade that young man to accept a position Millikan had finally offered—the noun or its adjective was constantly evoked.

"Advancement," he wrote, "will depend first on productivity." In a later letter he wrote, "Advancement will be more rapid than is ordinarily the case, provided only that you continue to be productive."

Bell knew as well as anyone that the quality of mathematical research cannot be measured by the number of papers published, or the number of pages; nevertheless, he had expected more from Robertson than he was seeing in print. The young man appeared to be spreading himself over too many different fields.

This last was to be a lifelong criticism. In 1951, when Robertson was finally elected to the National Academy of Sciences, Bell brought up the subject again.

"This [election to the Academy] is something many years overdue, mainly your own fault, for straddling the fence between two sciences and not letting your fellows know on which side your balls hung. For it is impossible to part that pair."

Robertson's bibliography in the six years between 1924 to 1929 included differential geometry, the theory of continuous groups, atomic and quantum physics, and general-relativistic cosmology, according to the astrophysicist Jesse L. Greenstein, who wrote the Robertson biographical memoir for the National Academy of Sciences; but there had been only seven papers published in six years. Young Michal, on the other hand, had been commendably productive at Ohio State and in the two years he had been there had also seen two candidates through the Ph.D. Bell pressed hard for Michal to accept Millikan's offer of $4500, "grade to be discussed."

But Michal wanted a guarantee of $5000 in two years.

"I turned down a $7500 research professorship ... to come here at $6000," Bell told him. "I asked for no raise, and was promised none .... I am backing you to make good, and I have told Millikan so, on two grounds: (1) that you work hard and are interested in research, and that you will most likely keep your ambitions and continue to be a producer; (2) that you are trained in a modern field ...."

Finally Michal accepted a second offer from Millikan, which gave him the rank of associate professor. The day after Bell heard about Michal's acceptance he wrote to Robertson, expressing (it appears) his disappointment in Robertson's work during the past two years. He may also have been reacting to an undated telegram from Robertson to Millikan, sent without any word to himself, in which that young man proposed extending his leave of absence from Caltech for another year so that he could accept an offer from Princeton, where he would have contact with Einstein. Whatever Bell actually wrote and whatever his motivation, Robertson was deeply and permanently hurt by Bell's letter, according to his daughter, Marietta Fay.

There are several later notes to and from Bell among Robertson's papers in the archives at Caltech. They are goodhumored and usually raucous although—I may imagine it—there always seems to be a little more reserve on Robertson's part than on Bell's. Nevertheless both men invariably closed their communications with a secret expression from the days of their early friendship, *AHYBASI!* The letter that Bell wrote to Robertson on March 20, 1929, is not among the papers in the archives. Marietta Fay tells me that she remembers seeing it, and she knows very well the gist of it—in fact, she was once going to frame it with the inscription "Publish, or Perish." The letter is essentially authenticated by Robertson's response, which he directed to Millikan rather than to Bell. In it he asked to be released from the appointment he had accepted with such pleasure two years earlier. Robertson's request was granted without, apparently, any further correspondence. He went to Princeton as an assistant professor and spent much of the next two years translating Hermann Weyl's classic work on group theory and quantum mechanics.

"Both my parents were deeply hurt by Bell's letter," Marietta Fay tells me. "They never *forgave* him."

I object that when I ask people, "Who were Bell's friends?" they all say, "The Robertsons." In letters Bell wrote to Taine after the Second World War, he frequently wrote of "Hot Dog" being in Washington again and "leaving his family in my care." In one letter he wrote that Angela Robertson had offered to sit up all night with him when he was sick and that "Hot Dog comes in every day."

"*Of course,*" says Mrs. Fay. "They *would* do that. They were *friends.* But when my father finally came to the Institute in 1947 he made it a condition of his coming that he would not be a professor of mathematics, he would be a professor of physics."

Thus, instead of "Hot Diggety Dog" Robertson, it was Aristotle Demetrius Michal ("the Greek" to Bell) who arrived in August 1929, accompanied by his beautiful young wife, Luddye, to take up his duties on the mathematical faculty. Michal—"Aris" to his wife and "A. D." most often to colleagues and students—has been described by Angus Taylor, one of his Ph.D.'s at Caltech, as a small, unusually white-skinned man with a shock of very dark hair and a quick smile— enthusiastic and generous. "I don't know how he ever had any money when he died!" Sixteen years younger than Bell, he was a mathematician of another generation, much influenced by the new abstract point of view in mathematics. His wife, Luddye, was a charming, sweet-tempered young southern woman who was "very sharp and always interested in new ideas." At Ohio State she had been a member of a book club of faculty wives, and at Caltech she was instrumental in starting another such group. Known as "The Worms," it still exists as an active group. Her beauty was immediately adopted by the all-male student body at Caltech as a unit by which to measure feminine pulchritude.

"A 1.0 Michal was sheer perfection," I am told, "but a .8 Michal was still pretty good."

It was fortunate for Bell that Michal arrived when he did to help with the mathematics, for almost immediately Toby fell ill, the first real illness she had suffered in her life. Bell was distraught. He wrote in detail to friends in Seattle of the suffering of "Dog Toby, my wife." At first the doctor had thought that it was cancer, but by the middle of October he had changed his diagnosis.

"I thank whatever gods may be for a merciful deliverance."

Then, just as things seemed to be going well, there was "a ghastly setback," and Bell had to arrange not to be away from the house for more than a couple of hours at a time. Not until December did he feel confident that she was going to get well.

Toby's illness coincided with the period immediately before and after the stock market crash of October 1929, but Bell made no mention of it in his letters. The next two years would see the country pass from prosperity to depression, but Dr. Bell and his alter ego, John Taine, would not be immediately affected. On the campus of the California Institute of Technology, Millikan's idea of persuading a hundred wealthy Southern Californians to contribute a thousand dollars apiece for the next ten years in return for being immortalized as "The Associates of the California Institute of Technology" had turned out to be another stroke of genius on his part. Early in 1929 Allan and Janet Balch, founding members of the Associates, had set aside stocks and bonds to build a facility on the campus (described as "the hearthstone of the Associates") that would serve the staffs of the Institute, the Huntington Library, and the Mt. Wilson Observatory as well as the Associates themselves. Those who had custody of the gift had—with remarkable foresight—converted the holdings to cash before October 1929. As a result the Institute had half a million dollars designated for that one building, on which work started at the beginning of 1930.

Again, even from these years, there are not many people still alive who remember Bell at Caltech. However, it is my good fortune that shortly before Carl Anderson's death, Laura Marcus interviewed him about his memories of Bell. Anderson took his Ph.D. in 1930, having already begun the work for which six years later he would receive the Nobel Prize in Physics. Since he minored in mathematics, he had to take Bell's course in Modern Analysis. He found him poorly prepared but the course interesting "because he was such an interesting person." He repeated the phrase "such an interesting person" several times to Mrs. Marcus.

Bell was also a member of Anderson's examining committee.

"So at nine in the morning I was there and so was Dr. Bell, but no one else. He asked me some questions for about twenty minutes. But still no one else was there. He was very much interested in the history of mathematics, and so for about half an hour during my Ph.D. exam I sat there listening to Dr. Bell talk about Bessel's childhood."

In the context of my search for the man who was both E. T. Bell and John Taine, the years covered in this chapter are unusual. Although they are lacking in personal recollections, they are treated in two sets of personally revealing correspondence. In addition to Bell's letters to

Michal, I have been fortunate to receive from the archives of Syracuse University copies of letters exchanged during this period between Bell and his publisher, E. P. Dutton and Company.

Having finished his Colloquium Lectures and the accompanying book, *Algebraic Arithmetic*, Bell returned to the writing of another scientific adventure yarn. "At this up-to-the-minute scientific research institute," he found "more plots for thrillers in a week than I could write out in a lifetime." His "fictive orgy" during the Christmas holidays of 1927–1928 produced a story that he titled *White Lily* after its beautiful Chinese heroine.* It begins with a mother and son dyeing Easter eggs while the father, a captain in the Marines, is on his way to China. What ensues is based on the premise, as Bell wrote to Macrae, the president of Dutton, "that it is not at all an improbability that chance might have tipped evolution in favor of silicon versus carbon." (The novel later appeared as *The Crystal Horde*.) He sent the manuscript off with two early narrative poems and two novels he had written while at Washington—*Red and Yellow* and *Desmond*, now retitled *Damon*.

At the beginning of June 1928, before he went to Chicago for his final summer session there, he completed still another novel. This one, he explained to Dutton, was about "the very latest stuff in radiology as applied to Evolution." To support his statement, he enclosed two clippings from *Science*: "One relates to the new work on the artificial production of mutations (permanent) in living animals by x-ray work; the other to our own x-ray work at the Institute." This was work "which will have far reaching consequences upon the future of both sciences, and possibly upon human life in the most intimate sense . . . ." The laboratory in his book was too close to T. H. Morgan's laboratory in Pasadena so he had set the story in Seattle.

"For the first time in my stories I have created a superman," he told Dutton. "Unlike most supermen in literature, this fellow . . . does think and act as a real superman might reasonably be expected to do."

The correspondence between Bell and the people at Dutton reveals what must have been a very frustrating lag in that publisher's response to manuscripts he sent. Finally he wrote his editor that his last manuscript (titled *Against Himself* but ultimately *Seeds of Life*) should have "rather prompt" attention.

---

* The holograph manuscript in the Institute Archives ("Written at Caltech at Christmas 1928–1929") is misdated, since Bell refers to the novel in letters to Dutton before that date.

"It is the first novel of its kind in the field. The scientific background in it is new; not even Wells, the professed biological scientific romanticist, ever had an inkling of it. Tomorrow the idea will be as common as dirt."

If Dutton wasn't interested, he would try to place the novel elsewhere. Within a week he had an answer. Since *Green Fire* had just appeared, it seemed best to wait until spring to publish *The Greatest Adventure* or *The Iron Star*—the two still unpublished novels for which Dutton had contracted. Within the next two weeks he would be hearing from Mr. Macrae himself about the handling of his future work.

Macrae's letter (two single-spaced typewritten pages dated November 2, 1928) is clearly the result of careful thought. *Desmond* was well written, but even as *Damon* it was still unpublishable because of its closeness to the Lodge story. (It illustrates in its various versions Bell's inability or his unwillingness to work over a manuscript. If a story did not come easily, he once told Durst, he simply threw it away.) Both *White Lily* and *Red and Yellow* were below Bell's standard. After *Green Fire* they would be "an anticlimax." As for *Seeds of Life*—in organization, dialogue, and dramatic effects it represented a "highwater mark in [Bell's] library of scientific romances." It was a very original presentation, Macrae wrote, "but after having read [it] I cannot rid myself of a certain feeling of revulsion that a human being should be subjected to so monstrous and abhorrent an experiment even in the interest of science." Macrae then moved on to the future handling of Bell's novels.

"I have always expressed an admiration of your work," he told Bell, "and I have held the belief that you were going to write a book which the American public would like and buy in large numbers, if for no other reason than the sheer originality of your concepts. Thus far my ambition for you has not been fulfilled and the financial loss to me has been a matter of serious concern. I have gone over the situation in my mind and I have come to the conclusion that it is not the fault of the material but the form in which it is presented, and this suggests to me that perhaps you could give me more salable novels if you were to have an experienced collaborator who could join you in the actual writing of them."

Bell responded in the same frank and friendly tone. He made several suggestions that might make *Seeds of Life* more palatable. He was not opposed to making changes but—as to collaboration—he was unwilling to accept anything other than the kind of editorial help he had received on *Green Fire,* "[which] I consider remarkable." Collaboration "in its full and usual sense of joint authorship" he rejected out of hand.

"I feel that the scientific ideas in these stories are almost certainly beyond the initiative of any ordinary writer who has not had the advantage of a minute and extensive knowledge of a good deal of current science. It may not be apparent in these books to the casual reader that there is creative thought in them; this can be contributed only by someone whose life work is scientific research. It cannot be too strongly emphasized that a coherent, scientific imaginativeness is not possible to one who is not primarily a scientist. Therefore, I believe that these stories, whether good or bad, could not be written by a purely literary man, no matter how gifted. . . . [B]arring ordinary human mistakes, my training enables me to write stories that, no matter how wild apparently, are logically selfconsistent. They do not fall apart; within their own data they hang together."

Bell had clearly become aware of the growing number of magazines devoted to "science fiction," for he now proposed to Macrae that it might be "good advertising" if the books, although different from those published by Dutton, were to appear—he even mentioned the possibility of serialization—"in some medium which reaches a different stratum of readers than the extremely sophisticated one to which your house has always catered."

Six weeks after Bell wrote the above letter to Macrae, he again felt the urge to take up his pencil. On December 4, 1928, he began *The Forbidden Garden,* which he completed on January 16, 1929. The holograph manuscript is among the Lloyd Eshbach Papers at Temple University with the dates in Bell's hand on the first and last pages. In Chapter XIV he took note of the fact that he was writing in San Diego and in Chapter XVI, that the date on which he was writing was January 4. It was the thirty-third anniversary of the death of his father.

The Forbidden Garden, even in its title, is a psychologically interesting story to anyone familiar with Bell's family history. The brothers James and Charles Brassey come from a family with a genetic flaw that has marked one member of each generation with insanity. Charles now heads Brassey House, a famous English seed firm, while the "tainted" James has voluntarily exiled himself to India. Although Bell liked to play games in naming his characters, one cannot help being struck by the fact that he has given the older, "tainted" brother the first name of his own older brother in Ceylon. James Brassey is described as having "gone native," but with his "inherited and cultivated love of all growing things" he has sent back to his younger brother some previously unknown seeds. The only one that has germinated has produced an unusual but sterile flower of a fabulous blue. The "memory source" for this exotic flower appears to be in a poem by Bell in the archives at Santa Cruz. Written during his early years at Washington, it describes "myself at four years old/ Beneath a tree I stand/ A bluebell in my hand/ And far away the hills . . ." The poet is clearly back in the paradise of the Santa Clara Valley where "The bluebell never fades."

The expedition that sets out to bring back some of the uranium-rich soil in which the exotic blue flower originally grew is an assorted group, none of whom is what he or she appears to be. The "native guide" is initially unmasked as Jamieson (a suggestive name that Bell uses in other novels), the head of the Indian Secret Service, who has been working for ten years with the head of Scotland Yard to find the spies that Charles is convinced have infiltrated Brassey House to learn the secret of the blue flower.

Bell is ambivalent in his treatment of his principal characters. At the beginning the reader is told that James is the brother who has inherited the family insanity, which is characterized by paranoia. Later he is told that it is Charles who is insane—and Charles does indeed seem paranoid on the subject of spies at Brassey house. James talks sanely to his visitors and then slides into insanity, destroying himself and his kingdom. At the end of the book the head of Scotland yard announces that Charles Brassey is "the sanest man I know." Jamieson, when he is first unmasked, is scorned by the others as "not English" and thus lacking "the sporting sense," but he stands up to them and defends his mixed heritage: "I am proud that my mother's mother was a full-blooded Indian woman, prouder far than of the fact that my mother's father was an English officer, although I am proud of that too." Unmasked again, as a member of a fanatic organization that hopes to finance its activities by obtaining the precious soil, he is physically and verbally bullied by another member of the expedition.

Bell did not send *The Forbidden Garden* to Dutton until the summer of 1929, noting in a postscript to his cover letter: "This novel deals with data which are known only to those scientists in immediate touch with current research; but these things will be common property tomorrow." *Garden*, however, was not to be published until 1947, and then not by Dutton.

The new John Taine novel that Dutton published in 1929 was *The Greatest Adventure*. A brief review, merely summarizing the plot, appeared in Gernsback's *Amazing Stories*—the first notice that the new scientifiction press (as it was then called) seems to have taken of John Taine. Bleiler, in his summaries of science fiction of the 1920s, notes that it is generally considered one of John Taine's better novels in this group. Under the heading "E. T. Bell's/Slimy Monsters/Bloodcurdling," the novel was also reviewed—another first—in the *California Tech*. According to the student reviewer, *The Greatest Adventure* gave "the general impression [of] a mixture of H. Rider Haggard, Conan Doyle, Roy Chapman Andrews, and a bottle of excellent gin." (This last was, as Toby would have said, a slander. Bell's drink was bourbon.)

Following a night spent on Mt. Wilson watching Edwin Hubble at work, Bell next wrote—probably in the summer of 1929—what he described as "a scientific novel with a biological and astrophysical background." He called it *Tomorrow* and sent it to Dutton on November 25, 1929. He still had lots of ideas for novels, he told the publisher. There were plots of fourteen or more "pretty fully outlined" in his head. One he described as "a sequel to *Green Fire*, in which there is a real 'war of the worlds,' on strictly modern scientific principles."

As usual Macrae did not reply promptly. Bell—he finally wrote—had "a most remarkable and marvelous imagination." He (Macrae) had personally read and enjoyed all the John Taine books, but he had also lost money on all of them except *The Purple Sapphire*. It was a pity that Bell could not be prevailed upon to collaborate with someone who could work his ideas into a scientific mystery story "which we can induce the public to buy . . . ." He returned the manuscript of *Tomorrow*.

As sometimes happens, the closing of one door is the opening of another. The publication of *The Iron Star* in January 1930, the last book by John Taine that Dutton published, coincided with the first appearance of John Taine in a science fiction magazine. The winter issue of *Amazing Stories Quarterly* announced on its cover "A Complete Novel of Surpassing Interest." This was *White Lily*, rejected by Dutton as not up to standard.

Response to the novel, which was issued later as *The Crystal Horde*, was varied. Professor Campbell found it "artistically one of the most notable of [John Taine's] works," of all his novels "best combin[ing] bold fantasy with finely realized fictional verisimilitude." Groff Conklin, reviewing the book for *Galaxy*, described it as "contain[ing] one of the most magnificent sci-horror ideas ever created in the Earth-cataclysm genre, and probably the worst yellow-menace plus Bolsheviks plus religious prejudice melange ever to hit science fiction."

The year 1930 was a significant one for both Mr. Taine and Dr. Bell. It saw the end of John Taine's association with Dutton and the beginning of his career in the new science fiction press. That same year the name E. T. Bell appeared over fifteen published research papers, the largest number in any year of his mathematical career.

THE TIME STREAM

By the Author of "The Purple Sapphire," "Quayle's Invention," "The Iron Star," etc., etc.

CHAPTER I
Sent Back

WE have explored to its remotest wildernesses a region that all but a few hold to be inaccessible to the human mind. Yet, in looking back after twenty-five years at the colossal drama which unrolled with stunning rapidity before our bewildered consciousness, I can see in it all no incident more mysterious than the unquestioning faith with which we accepted our guide—the venerable Georges Savadan—at his own valuation. That granted, the rest followed with a magnificent inevitability. It was like the running down of Sylvester's watch when the mainspring snapped; a trivial accident precipitated events which time had been holding in suspension for ages.

We fell in with Savadan's suggestions as readily as if they had been the natural promptings of our own minds. Indeed, with the exception of Beckford, we seemed at every step of our progress into the unknown to anticipate the old man's will. The sudden transformation of the Savadan whom we had so often watched dreaming over his port at Holst's, into the resolute, energetic leader alert to every hint of danger, caused us no surprise. It was a change for which we were secretly prepared. We had known the old man subconsciously all our lives.

Surely there never was a company of explorers less likely than ours to penetrate the dim secrecies of the future; but so we did. It matters little what our occupations were when we set out, exactly a quarter of a century ago, on our explorations. Nevertheless, as our lives and all of our daily activities acquired a strange significance in the light of our adventures, I shall briefly state who and what we were, and how we started.

The place and date are extremely important. We first succeeded in entering the time stream in San Francisco, on the fourteenth of April, 1906. That was precisely four days before the city was destroyed by earthquake and fire.

The mere catalogue of our party follows, with their nationalities and places of birth, and their ages in 1906.

Colonel Dill, born in Tennessee, veteran of the Civil War, age 66. Habitually in need of money, and eager to imbibe any amount of whiskey at any other man's expense. Hobby, imaginary bloodshed.

Palgrave, born in San Francisco, age 26. Physician, specializing on morally and mentally unstable children. This man is now the well known child specialist of New York. Hobby, color analysis.

Beckford, also born in San Francisco, age 27. Attorney and orator. Lifelong friend of Palgrave and, with him, suitor of Cheryl Ainsworth. Hobby, oratory.

Ducasse, born in France, age 25. Psychologist and student of philosophy, with strong leaning toward practical mechanics. Hobby, prison reform.

Herron, born in Chicago, age 24. Newspaper reporter. Hobby, deciphering of code messages.

Culman, born in Germany, age 47. Mechanical engineer and inventor. In easy circumstances, owing to a simple, lucky device for measuring flow in oil wells. Hobbies, hatred of war and a desire to expose the trickeries of mediums and others in some kinds of psychic research.

Sylvester, born in England, naturalized U. S. citizen, age 23. Lived at Los Gatos (three hours by train from San Francisco), where he had a paying ranch. Highly educated in modern theoretical physics—as it was in 1906, with ample leisure to continue his studies.

Savadan, born in France, age unknown, probably 65. Political refugee. Had lived for many years in San Francisco in very straightened circumstances.

Myself (Smith), born in San Francisco, age 24. Analytical chemist.

To this strange assortment three others may be added, although they were not strictly in our party. First there was Cheryl Ainsworth, age 24, who

JOHN TAINE

THE idea of travel in time, has no doubt captured the imaginations of our readers. Discussions pro and con, as to the nature of time, and the possibilities and effect of travel in time have waged through our discussion columns and an enormous amount of interest has been stirred.

What could be better to clarify that interest than a story by a man who is not only a master of science fiction, but also a scientist of the first rank. John Taine, is the pseudonym of a professor of mathematics at a leading American University, and a member of the National Academy of Science, that most aristocratic body of our foremost thinkers.

If the number of books of science fiction in print is any criterion then John Taine ranks with H. G. Wells. His "Purple Sapphire" created a virtual sensation. "Quayle's Invention," "The Iron Star," "Green Fire," "The Gold Tooth," "The Greatest Adventure," have been no less of successes. To this great company of novels, "The Time Stream" must now be added.

John Taine has probed deeply into the nature of time, and of time travel, and his ingenious mind has created a tremendous and thrilling story of the whole human race. The rise and fall of the race; the great cataclysm that sent men back to the brute; are all woven into a powerful and moving narrative. We are honored to present to our readers this science fiction classic.

THE TIME STREAM (1931) with note by Bell as to the date of writing

Queen of the Sciences

By September 1930, as Bell began his fifth year at Caltech, the economic depression of which the stock market crash had been the precursor was widening and deepening. In this situation an optimistic group of men nevertheless began to plan an exposition to celebrate Chicago's centenary. "The Century of Progress" would be "A Nationwide Adventure in Education"—its theme, "the forces of science linked with the forces of industry."

This theme had been the suggestion of the distinguished inventor, M. I. Pupin, upon whom Bell had based the mad scientist of *Green Fire* (currently being dramatized by his friend Glenn Hughes). Taking up Pupin's suggestion, those organizing the exposition had appealed to the National Research Council "to devise a plan of exhibits by which the story of the sciences could be told in its entirety, and yet swiftly and with a simplicity of detail that would make it clear and absorbingly interesting to everyone." The Council responded, not only by devising such a plan, but also by suggesting the publication of a series of popular little books on the sciences and various industries. Such books, it suggested, could reach an audience much greater than the thousands who would be able to come to Chicago in the summer of 1933. The idea was so similar to the plan for cheap and nontechnical little books about science that Bell had proposed at the end of his still unpublished novel *Desmond* that when I first come across it I wonder if it might not have been his suggestion; however, I uncover nothing to indicate that it was.

The years between 1930 and 1933, when the Century of Progress was coming into being, are better documented than are most years in

Bell's career thanks to letters from him (and on occasion from Toby) to Glenn Hughes, by then Professor of English and Head of the Division of Dramatic Art at Washington. In addition, Hughes's profile of Bell, published in October 1930, provides a contemporary description: "a slender, wiry individual of medium height, with clear-cut aquiline features, hair graying at the temples, and penetrating but extremely kind eyes." I must admit that at this point in my search for Bell the word *kind* applied to him comes as something of a surprise. I am, however, to hear the adjective with increasing frequency as, farther up the time line of his life, I find more and more people who knew him.

Bell's letters to Hughes (there are none from Hughes) are obviously letters to a good friend, but the occasion for writing is usually business—most frequently decisions regarding the production of *Green Fire* at the Pasadena Community Playhouse, possibilities of interesting "movie people" in that novel and others, disagreements over royalties with Dutton.

In 1930–1931 Bell was complaining to Hughes that he was "being driven crazy" by the four NRC Fellows in mathematics who were on campus that year. Actually he must have taken considerable satisfaction in their number and the fact that all four had come to Caltech because of his presence there. Among these were young Lehmer, later at UC Berkeley, and Leonard Carlitz, later at Duke University. It was to Carlitz that Bell once murmured, perhaps a little wistfully, that he would like to have some numbers in mathematics named after him.

In 1930 Lehmer had just received his Ph.D. from Brown, and Carlitz his degree from the University of Pennsylvania. Both men found Bell's seminar on number theory "a very high-class affair." Attendance was not large, ten or twelve at the most; but young Morgan Ward, although an instructor by then, attended as did some other established mathematicians from the wider Los Angeles area.

Since there had been no one at Brown who felt competent to pass on Lehmer's doctoral dissertation, it had been sent to Bell for approval. As a result Lehmer always considered himself in a sense a student of E. T. Bell. In his *Selected Papers* there is a group of seven papers concerned with the general algebra of numerical functions that are specifically cited as being inspired by Bell's seminar of 1930.

That fall, according to Lehmer, the vacant lot next to Bell's house was put up for sale. Bell was eager to buy the property, which he had been gardening as if it were his own, so "he holed up in San Jose for a week," and wrote a science fiction novel, the sale of which gave him

LEONARD CARLITZ
in Bell's 1930 seminar

enough money for the purchase. Later I see in the memorabilia at Watsonville a letter from Taine to his father addressed "In Care of General Delivery, San Jose"; but I cannot match Lehmer's recollection with the admittedly small amount of documentary evidence I have about the early sales of Bell's novels to the pulps. All the novels that were published in 1930 and 1931 had already been written by the end of 1929. It is possible that rather than writing a novel in a week (which would have been very fast even for "John Taine"), he may have devoted himself to shortening *The Time Stream* to make it more salable (something he had mentioned to Macrae). Certainly by this time he must have recognized that the readers of the new science fiction magazines were that "peculiar audience" that he had always thought his story of circular time and concurrent existences required. In fact, in the issue of *Amazing Stories Quarterly* in which *White Lily* appeared there was an editorial, which he could not have failed to see, on the three main themes in contemporary science fiction: time travel, space travel, and utopias.

At the going rate of one cent a word it seems hardly possible that the sale of a novel could have enabled Bell to buy a large corner lot in Pasadena, even in the Great Depression. But he was not opposed to buying "on the installment plan." The lot was purchased, and plans were made to build a small hexagonal "shack" on it that would serve him as an "outdoor study."

From Bell's letters to Hughes it is clear that in the fall of 1930 the Depression was still scarcely being felt in Pasadena. The sumptuous new "Athenaeum" had just been completed, and in one of Toby's letters to Hughes she excitedly described its luxurious amenities.* The facility did indeed become, as planned, a gathering place for local scholars and "visiting eminences" (so Toby described them). Economical as well as elegant, it housed even graduate students during the depression years.

Toby's letters to Hughes are much more informative than Bell's and enable me to place many of Taine's recollections of his family's life in Pasadena. From them I learn that in the 1930s the Bells finally decided to learn to drive. Toby, who was over fifty by that time, took to the wheel as if born to it, but Bell was physically awkward considerably beyond the handicap of the missing first joint of his right thumb. (In fact, the lack of a thumb was not much of a handicap, for he wrote swiftly and clearly with his right hand.) When he failed his third driving test by running over a curb, he was dreadfully embarrassed. He gave up trying to learn to drive and instead bought Toby a "snappy" black Ford roadster with red wheels and a khaki convertible top.

"Have you seen the model?" she demanded of Hughes. "His" name was Archimedes, and "he" was very intelligent. "Whenever I do anything wrong, he stalls, which is safe though slightly embarrassing." She had shocked people by immediately taking him into downtown Los Angeles. "It seems that isn't done unless you have driven for a long time. I don't see why, but so many of these conventions are difficult to explain."

The roadster became a familiar sight in Pasadena, Toby driving, smoking a long thin cigar and wearing a big floppy hat, Bell most often curled up in back in the rumble seat.

In the same letter in which Toby described "Archimedes," Bell acknowledged the receipt of a copy of his short essay, *Debunking Science*, which was No. 44 in Hughes's *University of Washington Chapbooks*. These were pamphlets appearing once a month and usually featuring

---

* At Caltech the choice of name has been explained as follows: "In Athens, at a temple named in her honor, The Athenaeum, the city's stellar intellects gathered to hold forth on their latest ideas. In time, Rome eclipsed Athens [and] when the Emperor Hadrian built an academy in Rome, he took Athena's name for his building. In time, Pasadena eclipsed Rome . . . ."

well-known names. The subject of Bell's essay was the ongoing and heated debate in mathematical circles on "The Crisis in the Foundations." It was a subject that some mathematicians did not even mention in nonmathematical circles, fearing that laymen might get the idea that mathematics was no longer reliable. But Bell did not hesitate.

"The one thing that has become increasingly clear above all others in the past twenty years is this: there is at present no certainty whatever regarding the major conclusions of modern pure mathematics, nor is there any immediate prospect that pure mathematics as it exists today will be proved to be self-consistent."

The work of debunking science as a whole—of finding out exactly how much could be safely trusted—had begun in the revision of mathematics itself— the queen of the sciences—by those calling themselves finitists or intuitionists and led by the Dutch mathematician L. E. J. Brouwer. Their program to put mathematics on an unquestionably sound basis would eliminate vast areas of the subject that involved the hard-to-manage infinite and would eliminate as well the Aristotelian "law of the excluded middle"—that something must either be or not be, that there could be no third possibility. The elimination of this law from permissible mathematical reasoning would in turn do away with so-called "existence proofs"—a treasured tool of mathematicians since the time of the Greeks.

Opposing Brouwer—defending, as he said, "the paradise of the infinite" that Cantor had created and insisting that depriving a mathematician of existence proofs would be "like depriving a boxer of his gloves"—was David Hilbert, whom Bell described as "the dean of living mathematicians ... a man whose tireless genius for nearly fifty years has enriched mathematics with dozens of beautiful things." Hilbert hoped that his Theory of Demonstration would establish with finality—Bell's words—"that the mathematical game, played according to certain innocent enough rules" was of necessity *consistent*: in short, that these rules *could not lead* to a proof that a theorem was true— as well as a proof that it was not true.

Reviewing *Debunking Science* in the *American Mathematical Monthly*, Mark H. Ingraham provided his own table of contents in the spirit in which Bell had written his essay:

Part I. Stop Being Childish and Dishonest ... (This part is a pleasant vituperation against credulity and is very good reading ....)

Part II. Even Mathematics Needs Debunking ... (This part is a little more solemn. Toe treading isn't so much fun among friends. This chapter, however, gives a very good account of the highly optimistic attitude of pre-Brouwer mathematics.)

Part III. Description of the Mathematical Controversy ... (This is really a very good non-technical hint of what is going on in mathematics. It is about all that can accurately be said to the non-mathematician. Although well written, it will never be included in the "World's Greatest Humor.")

Bell had enjoyed writing *Debunking Science*, and he now wrote a similar but more wide-ranging essay on "Science and Speculation." The voice in both of these efforts was clearly not that of J. T. nor of John Taine. It was that of E. T. Bell, writing under the name that appeared above his mathematical research but not—by the farthest stretch of the imagination—writing like a mathematician.

Debunking Science was beginning to circulate and "Science and Speculation" was scheduled to appear in *Scientific Monthly* when, at the end of 1930, Bell was elected president of the Mathematical Association of America with 330 votes to Graustein's 255. His election was the culmination of his activity in various scientific organizations—activity that had begun during his last years at Washington.

"'Honors,' if they can be called such," he noted on a form he filled out for the administration at Caltech, "most have involved a lot of hard, unpaid work."

On the basis of his letters in the MAA archives at the University of Texas, Bell's presidency appears to have been more or less routine. He suggested that the Association might broaden its membership by canvassing the junior colleges; but W. D. Cairns, the Secretary, pointed out that it was more difficult to obtain the names of junior college mathematics teachers than the names of those at colleges and universities. Bell then urged that an effort be made to improve the mathematical quality of talks at meetings of the Association.

"I believe in papers that can be understood by members ... but I think we ought to draw the line somewhere. The only thing I can think of in the present connection is to shoot the program committee."

Cairns noted on the letter that this complaint should be discounted a little "on the score of Bell's drastic assertions."

The only letter of general interest in Bell's official MAA correspondence has to do with the Association's suggestions for European mathematicians to be invited to the Century of Progress Exposition. Writing on January 26, 1931, again to Cairns, Bell was as usual emphatic.

"I think the committee will overlook a good bet if it ignores the thing of liveliest mathematical interest today, namely the controversy over the foundations. If Hilbert is still living, he should be invited; but above all others, Brouwer should be asked."

The principle, he went on, was simply "which of these men has started something really new?"* He felt that most of the mathematicians he vetoed were "no better than what we have ourselves, and in fact not so good. I think we should not cater on this occasion to the Europeans' already high opinion of themselves . . . . We don't have to kiss anybody's foot."†

At almost exactly the same time that Bell was writing this letter, he was asked to write 30,000 words on mathematics for the Century of Progress Series. Two months later, having been "tied up longer than necessary on some mathematics, chiefly due to interruptions"—those bothersome NRC Fellows again—he had finished the book. At some point he had changed the title to *Queen of the Sciences*, Gauss's famous description of mathematics.

"Whether as history or prophecy," Gauss's description was "far from an overstatement," he informed his readers on the opening page. Time after time in the past two centuries major theories in the physical sciences had come into being "only because the very ideas in terms of which the theories have meaning were created by mathematicians years, or decades, or even centuries before anyone foresaw possible applications to science." But these applications had not been the goals of the mathematicians whose theories had made them possible.

"Guided only by their feeling for symmetry, simplicity, and generality, and an indefinable sense of the fitness of things, creative mathematicians now as in the past are inspired by the art of mathematics rather than by any prospect of ultimate usefulness."

---

* He voted "a decided no" on Bernstein, Besicovitch, Carleman, Levi-Civita, Féjer, Haar, Hahn, Hausdorff, Herglotz, Julia, Lichtenstein, Nørlund, Ostrowski, Perron, Pólya, Riesz, Schur, Sommerfeld, Szász, Szegö, and Weyl. "A doubtful yes" to Artin, Fréchet, Titchmarsh, and Landau. He gave "a decided yes to Hardy, who beats Landau out of sight at his own game, and who also is far more mature than Titchmarsh, the same vote for Hadamard, and possibly von Neumann. The doubt on the last is that the man still seems to be tied to Hilbert's apron string. A strong vote for Harald Bohr. I imagine a strong vote for Tonelli, I would take Bliss's opinion on this."

† Bell was shortly to describe himself in an interview with Lee Shippey of the *Los Angeles Times* as "two hundred percent American."

Conceding in his opening pages that "it is a waste of time for those who are not mathematicians by trade to explore the minutiae of modern mathematics," he then proceeded to sweep through modern mathematics in nine relatively short chapters, beginning with "Mathematical Truth" and ending with "Bedrock," or mathematical existence, all the while casually throwing in the names and works of some sixty mathematicians. His book was the first in the Series to be received by the publishers.

The spring of 1931 was a busy, productive one for Bell. He was "deluged" with fan mail as a result of *Debunking Science* and "Science and Speculation." Shortly after he finished *Queen of the Sciences,* he sold *Against Himself* (now *Seeds of Life*) to *Amazing Stories Quarterly* and had a request, so he wrote Hughes, for a story from another magazine. This was probably from *Wonder Stories* for his novel "about time."

The Community Playhouse planned to produce Hughes's dramatization of *Green Fire.* The actor playing "Dr. Ferguson," the head of the "good" Scottish laboratory, was to be made up to resemble Millikan, and Mrs. Millikan had taken home a copy of the script to make sure that the character was "worthy" of her husband. The physicists were all "doing their darndest to make [the play] go." Fritz Zwicky, "one of the best men on our staff," had taken on the job of creating the requisite green fire. Bell was hoping that Bela Lugosi, fresh from a great success as Dracula, might play Jevic. Lugosi did not take the part, but motion picture people seemed interested in the play and some of Bell's other novels.

On May 29, 1931, the *Pasadena Post* announced:

'Green Fire' Dazzles Throng
at Playhouse; Brilliant Cast,
Phenomenal Stage Settings

That October *Amazing Stories Quarterly* published *Seeds of Life,* the superman novel that Macrae had found "repulsive." Bell had named his protagonist "Neil Bork," a play on the name of the admired and much loved physicist Niels Bohr. Apparently, as in the case of Pupin/ Jevic, Bell liked to look at an outstanding scientist and ask himself how things might have been different. What if, in Bohr's case, instead of being born the genius that he was, he had evolved from some inferior human being? Bell's "Neil Bork" is a bumbling alcoholic whose clumsiness in the laboratory causes an accident that results in his evolving into a superman. Arriving back on the scene as Miguel De Soto, a Bolivian physicist, he guides Andrew Crane's laboratory to achievements Crane realizes he could never even have conceived. He also wins the hand of Alice Kent, whom Crane loves. Then suddenly he begins to lose his powers. Evolution is reversing. Alice dies giving birth to a monstrous reptilian son. The end of the story is bleak. For one generation the females of the human race will bring forth their reptilian young alive. Then the pulse of cosmic rays, generated from the disintegration of universal matter, will sterilize human and reptile alike. According to a note by Bell on a bound copy of the story, the last paragraph, in which DeSoto and his offspring are killed, "was substituted by the lady editor for the original ending, which the E. said made her sick."

I am interested in what Hughes, a professor of English literature, thought of Bell's novels.

"Most fiction writers are, after all, primarily fiction writers, and their excursions into exact sciences are hasty," he wrote of Bell. "Some of them may show a trifle more finesse in plot handling or characterization, but none of them surpasses Bell in grandness of conception or accuracy of detail. One has always the uncanny feeling that [he] is dealing in probabilities, and that many of his most extravagant dreams are but pre-visions of nightmares in store for the human race."

The month after John Taine's *Seeds of Life* appeared as a complete novel in *Amazing Stories Quarterly,* Robert S. Gill of the Williams & Wilkins Company circulated a letter announcing the publication of *The Queen of the Sciences* by E. T. Bell.

"Risk a dollar and get a copy," Gill urged. It would be worth that much in his opinion to learn about such an unusual endeavor in scientific education, but the purchaser would also get "about $5.00 worth of solid satisfaction" from Bell's book.

"It is surprising how comprehensively Dr. Bell has presented the fundamental concepts of mathematics in a short non-technical treatise," wrote the reviewer for the *American Mathematical Monthly.* Although the style was "mathematically unconventional," it would be "an unimaginative person who would not, from even a cursory reading, catch some of the glorious spirit of mathematics."

Bell thought, so he wrote to Hughes, that the limitation as to length had made *Queen of the Sciences* a better book than it might otherwise have been. Perhaps, Hughes suggested, Bell should consider teaching a general course in science. Bell had always believed in such a course, and he thought it could be "rather a joke" to teach, "as the lecturing would provide innumerable chances to turn out careful books on popular science." He had found that "without some such urge" it was "rather a bore" to do a popular book.

Queen of the Sciences sold quickly to the general public as well as to the mathematical community. It had been expected by the publishers that, mathematics being so unpopular a subject, it would come in last in sales; however, it "ousted the next 'best seller' by more than 3 to 1," Bell later reported.

In December 1931, the same month that E. T. Bell's *Queen of the Sciences* appeared, the first installment of John Taine's *Remembered Worlds* was published in *Wonder Stories* under the new and felicitous title, *The Time Stream*. But John Taine's identity was no longer being kept a secret. Each installment in *Wonder Stories* carried a sketch of the author and identified him as Prof. E. T. Bell of the California Institute of Technology.

By that time, ten years after Bell wrote the novel, the subject of time and time travel had become of great interest to readers of science fiction, but since the novel appeared in a magazine rather than as a book there were no "reviews." There were a number of comments in letters from readers in subsequent issues: "great work," "absolutely stupendous," "a landmark," "a classic in science fiction." P. Schuyler Miller, later the longtime book review editor of *Astounding Science Fiction*, wrote that "it requires plenty of rereading to really get at the meat. I know." The secretary of a chapter of the English Science Fiction League confessed, "I started off to write of Taine's *Time Stream* and, as is my wont, I read it through once more to crystalize my antipathy. Now I can't find anything wrong with it."

One subscriber, however, canceled his subscription because of a reference to "Hebrew" psychology: "I refuse to accept the nonsense of an Arabic mathematics, a French chemistry, and so forth."

The Editor wrote placatingly but did not improve matters by totally missing the reader's point: "Now as a matter of fact what Mr. Taine meant when he spoke of 'Hebrew psychology' was not all psychology, but the psychology of the 'unconscious' elaborated by Freud, Jung, Adler and others."

Throughout 1931 there was still hardly any mention in Bell's letters to Hughes of the worsening economic conditions. In the spring of that year—which he dismissed as "about the worst of the depression" —he received another raise of a thousand dollars:

"They ask us here not to discuss our salaries.* There is one thing however which is open news: they pay men according to their productivity regardless of rank or age. We have had an extremely productive year in mathematics."

Life for the Bells was still "a mad whirl of writing, painting, tending visiting eminences," as Toby described it to Hughes, "with the Dog Toby trying to attend to the business end."

"The best thing about this sort of life is the opportunity it gives for meeting the men who count in this scientific age and seeing them in action," Bell wrote in his 1931 letter to the *Eagle*. Einstein had just left the campus, and he thought the boys at his old school might be interested in his impressions of the most famous scientist in the world.

"The first time he was to talk formally, he came into the small lecture room and chatted about nothing in particular with his friends. Then, when he began to lecture, his whole demeanor changed. I can only describe it as entirely impersonal. It was as if a door had been opened in the sky, and a flood of light was pouring through. Einstein was the door; what he was saying was the light, and it did not seem to be his; it seemed to come from somewhere outside and beyond him. He lectured for two hours. . . . Then, suddenly, the light went out, and the door closed. Einstein again chatted with his friends or answered questions, an ordinary human being."

Of all the important men he had met, only two in his opinion—Einstein and the English number theorist G. H. Hardy, who had also visited Caltech—could be called geniuses. Outwardly they seemed completely different, and yet "in complete mastery of their stuff" he found them very much the same—"and there is no thrill quite so great to an American as that of seeing in action a man who is absolute master of his trade."

By the time, however, that the fifth and final installment of *The Time Stream* appeared in February 1932, the Great Depression was being felt even in the paradise of Pasadena.

"For the past week we have been in session here trying to meet a very serious deficit in our budget," Bell warned Hughes, who was interested in joining him at Caltech. "Drastic cuts have been made everywhere; these do not yet affect salaries."

---

* The information about salary came in response to a letter from Hughes in November 1931 reporting that Washington was interested in having Bell return to build up its mathematics department.

It was only a month later that Bell was reporting that Millikan had had to ask his faculty to take a voluntary ten percent cut. Severe as this may sound, it was nothing compared to what was happening at Whitman College in Walla Walla, Washington, where only the newly hired faculty members were being paid.

Bell continued to work, to teach, to do mathematics, and to write. Williams & Wilkins, pleased with the success of *Queen of the Sciences*, asked for another popular mathematical book and he decided to take as his subject numerology. He later described the book as "for the layman, and how!"

By then Hughes had had to cease publication of his Chapbook Series. As for Bell, although Williams & Wilkins signed a contract in August 1932 for *Numerology*, the book being by then completed, "they say in effect that the publishing business is so rotten at present that they will defer publication until 1933."

The Archives, California Institute of Technology

MATHEMATICAL FACULTY 1932: Front row, left to right, A. D. Michal, Harry Bateman, E. T. Bell, H. C. Van Buskirk; back row: W. M. Birchby, Morgan Ward, Unidentified, Luther Wear, Unidentified, R. S. Marlin, J. L. Botsford.

Throughout the early 1930s, in spite of his writing and his duties as president of the MAA, Bell was as ever extremely productive in terms of the number of mathematical papers that appeared under his name. He continued to exploit the main themes of the three areas of mathematical research on which he had concentrated at Washington and also began to produce in a fourth area: multiplicative diophantine analysis.

I have noted that the equations bearing the name of Diophantus of Alexandria were from the beginning of Bell's career objects of intense interest to him, almost all of his mathematical work being connected in one way or another with them. Now, however, he was taking up a larger and more general aspect of the subject, dealing not only with individual equations, but also with systems of such equations. By the end of 1932 he was looking forward to the publication of a very important paper in the *American Journal of Mathematics*.* "Reciprocal arrays and diophantine analysis" would mark the beginning of his work in the new field and, as Durst points out, "starts the work in this subject off with a big bang." In this one paper he presented his theory of multiplicative diophantine systems fully developed.

"The important point here," Durst emphasizes, "is that the methods applied to multiplicative diophantine systems are thoroughly general and completely devoid of the special kinds of arguments that had characterized so much of the work in this subject in the past."

During the same year (1933) Bell published three other papers on diophantine equations.

"Before Bell revolutionized parts of this subject sixty years ago, people relied on complicated arguments to find complete solutions for even simple diophantine equations," Durst points out. "Bell's main contribution was to simplify some of these arguments with a unifying principle he called 'The Basic Lemma of Multiplicative Diophantine Analysis,' the simplest example of the use of reciprocal arrays. It is called a lemma because the latter can be justified as a consequence of it."

For the general reader Bell's lemma (more felicitously in German a *Hilfsatz*, or "Help Theorem") is an example of his mathematical work that be easily grasped. It is simply stated:

---

* Bell's paper was published at the beginning of January 1933 in the same issue with Morgan Ward's paper "A type of multiplicative system." Between them, teacher and student, they had produced not one but two methods for finishing off an entire class of diophantine systems.

*The complete solution of a diophantine equation xy = uv is given by
x = ab, y = cd, u = ac, v = bd where a, b, c, d are arbitrary integers and b
and c have no common divisors.*

In itself the lemma is not an important result. What is important is what Bell showed could be accomplished with it. But once
again, Durst points out—as is so often the case with Bell's mathematical work—many modern textbooks in number theory seem unaware
of the lemma and its usefulness.

On the eve of the publication of "Reciprocal arrays and modern
diophantine analysis," at Christmas 1932, Bell suffered a heart attack.

I do not know how serious the attack was, but on March 29, 1933,
he wrote to Hughes that since Christmas he had been "living a strictly
godly, righteous & sober life." There was to be only one cup of coffee a
day and "absolutely no more alcohol." Dinner was to be eaten at noon.
("If you want to do something really disgusting, try eating in the middle of the day.") Unfortunately he had to confess that he had never felt
better in his life.

After the heart attack Bell begged off from personally delivering
his address as retiring President of the MAA on the grounds that crossing the Rockies was "above his height limit." Probably for the same reason he never attended the Century of Progress. Those who did travel
to Chicago in the next two years, and there were thousands who did,
found that the Hall of Science was the centerpiece of the Exposition.
Crossing to the Hall from the circular terrace and passing through the
pylons, they saw emblazoned above the entrance the words of Gauss,
and the title of Bell's little book on mathematics—*Queen of the Sciences.*

GRACE HUBBLE, frequently thanked by Bell for her part in his writings

E. T. BELL in 1933 at the midpoint of his academic career

Searching

In 1933 Bell was fifty years old and at the halfway mark in his academic career. It was essentially twenty years since he had received his Ph.D. from Columbia University. It would be another twenty before he would retire from the faculty of the California Institute of Technology.

At this point in his life I am able to talk to a few more people who knew him. The most helpful, because of his longtime contact, is R. P. Dilworth. Although he has been fighting a battle with cancer for many years, he generously agrees to talk with me on two occasions at his home overlooking Lake Oroville in the Sierra Nevada of Northern California.

Dilworth came to Caltech as a freshman in 1933, took his Ph.D. there in 1939, and became a member of the mathematical faculty in 1943. His earliest contact with Bell was as a member of the audience at a public lecture. The contact was more accurately with Mrs. Bell, who was sitting in front of him. Glancing around at the audience and recognizing a freshman, she promptly informed him that he was wasting his time—it would be a terrible lecture.

"And she was right," Dilworth laughs. "E. T. *was* a terrible lecturer, but at that time I was shocked that a wife would say something like that about her husband."

Later he learned that such blunt honesty was characteristic of Professor Bell's wife who, like Cheryl in *The Time Stream*, would have found it impossible (Bell wrote) "to write 'Yours sincerely' at the end of a letter unless she meant it."

Others who were at Caltech as graduate students and young instructors in the years immediately following Bell's fiftieth birthday are

Ivar Highberg (1933–1936), Angus Taylor (1933–1937), Donald Hyers (1934–1937), and Hubert Arnold (1934–1939). Most of their recollections are quite naturally of their own lives during the Great Depression when they subsisted on what they called "breadline fellowships" that provided room, board, and tuition without any money passing hands. All except Arnold recall taking Bell's course in Modern Algebra.

Algebra had changed greatly in the decade just past as a result of the work of Emmy Noether and her followers in Göttingen and of Emil Artin and Otto Schreier in Hamburg. What had been a very concrete subject had become very abstract. The new approach had been laid out in *Moderne Algebra* by B. L. van der Waerden, a member of Noether's group. Bell realized the importance of what had happened and felt that he should educate himself further in the new developments.

"But there is no getting around it, E. T. was not at home in abstraction," Dilworth tells me. "He had worked all his life with the integers. The natural numbers—1, 2, 3,. . . In mathematics you can't get much more concrete than the integers. He was just never at home in the new stuff."

The men I talk with have slightly varying recollections of Bell's Modern Algebra, but all agree that it was not a well-organized course. In Highberg's year the text was van der Waerden, which was in German. Bell's method of teaching was to read a sentence aloud and announce that he didn't believe it.

"By the time we students convinced him that it was true," concedes Highberg, "we pretty well understood it ourselves."

As times became harder, Bell abandoned van der Waerden, which was expensive, and returned to Dickson and, for the modern view, a little book by Helmut Hasse. Hyers recalls Bell's sitting in the back of the room and "picking away" at Dickson who, in Hyers's view, he looked upon as "kind of a rival."

Whether Bell actually felt this way about Dickson, who had played such an important role in gaining him national recognition, I cannot say. He spoke very highly of Dickson to H. A. Arnold.

"I remember his saying that there were very few people who had discovered even a typographical error in Gauss, but Dickson had actually discovered a *mistake*.* He also mentioned Dickson's work on

---

* An endnote in *The Development of Mathematics* refers to "the disastrous fallacy in Art.299 of the *Disquisitiones,* first detected by Dickson (1921), [which] retarded or misled diophantine analysis for 120 years."

Waring's problem as very deep. And of course he admired Dickson's history of the theory of numbers."

Bell did have a way of seeming at least to turn against people of whom he had apparently thought very highly in the past. Arnold and others tell me quite positively that Bell "didn't care much for" (some say "hated") President Suzzallo of the University of Washington. Yet I have difficulty making this statement jibe with the letter Bell wrote Suzzallo in 1925 when he turned down an offer from another university in favor of Washington, or with the supportive telegram he sent from Caltech in 1926 when he heard that Suzzallo had been fired: "It is nothing against you to be ousted by that pack of paretics . . . ."

I ask Dilworth how one can explain such conflicting behavior on Bell's part.

"That's E. T.," he replies. "That's all there is to it. He was, if you wish, a collection of conflicts. You have to take that into account."

I mention that often in reading one of Bell's articles, particularly on the crisis in foundations, I find it difficult to make up my mind as to which side of the argument he is supporting.

"That was certainly the case," Dilworth agrees. "It was hard to tell with him because he liked to be the eternal skeptic. I think it's probably true that you would be hard put to know, if you had to go down on one side of the fence or the other, where he stood. There are too many cases where I was having discussions with him in which he would be very skeptical and disparaging of something or other. He would make some very strong statements, and then a little later he would be perfectly—I am tempted to use the word 'reasonable' but maybe 'more temperate' is better. So there was this aspect with him that you were never quite sure, on any particular occasion, the position he was going to adopt."

"What *did* he stand for then?" I ask. "Is there anything on which you can say that there was no conflict?"

Dilworth thinks a moment.

"Well, there's no getting around it—and he wouldn't be skeptical about this at all—he certainly felt that the prime area of mathematics as far as he was concerned was the integers and their properties, and that finding out as much as you could about the structure and relationship of the integers was a major—and probably from his point of view *the* major goal of mathematics. I never heard him make a disparaging remark about number theory."

With the exception of Dilworth, the mathematicians I mention as being at Caltech in the 1930s were students of A. D. Michal—a fact that helps to explain the friction that was beginning to develop in the mathematics faculty. There was another factor. Michal was by then moving far beyond his teacher, Griffith Evans, in abstraction, "eagerly embrac[ing] the new functional analysis being developed by Fréchet, Banach, and others," according to Hyers. This was a development, as Taylor points out, that "afforded not only a generalization but also a simplification and a more fundamental approach to Michal's theories of differential geometry in function space." The new approach made it necessary to have research done on a number of different topics and required the collaboration of many students. Michal envisioned a "school" developing around himself.

Bell was doubtful about the very abstract and general ideas that were being so enthusiastically espoused by Michal. He was also disturbed by Michal's effective wooing of many of the best students that came to the Institute for graduate study in mathematics—in fact, it was rumored that Michal met the most promising students when they got off the train.

Bateman was not really involved in the competition.

"No one was his student in my time," Highberg recalls. "He was just too bright. The time would come when you had to select a topic for your dissertation. You would sit with him in his office, with his little four- by-six cards piled all over the table, and he would dream up a subject and roll it around in his mind. Then either he would solve it, or he couldn't. If he couldn't, you knew it was too tough for you."

But in spite of developing friction in the mathematics faculty, the Bells continued on friendly social terms with the Michals, because of the lovely Luddye.

When I telephone Mrs. Luddye Michal in Houston, where she now lives, she concedes that "Romps," who called her "the Yellow Rose of Texas," was indeed very fond of her.

"That was the name for General Santa Ana's mistress, you know," she explains in her gentle drawl. "But I don't think Romps actually knew who the Yellow Rose of Texas was."

When I ask about the professional friction between her husband and Bell, she says mildly, "Oh, there may have been a little of that from time to time, but you know how those things are. I think he [Bell] was quite different with men than he was with women. I remember he would say or do something to Aris, and Aris would say to me, 'He wouldn't be like that with you.' Sometimes there *would be* a big blowup, but then after a while everything would be fine and we would go out to dinner together again."

It is clear that Mrs. Michal was also very fond of Bell.

"Of course, he was always right on the ball. No matter what was said, he had an answer for it. He was that type, you know," she says. But she also recalls, both to Albers and to me, an occasion when she and her husband told Bell about the serious illness of the daughter of some friends of theirs— "and his eyes *filled with tears* at the thought of the suffering of that young girl, whom he did not even know."

At this time, in his fiftieth year, Bell was striking out in a new direction in his teaching. He did not hesitate to take up a subject outside his specialty. At Washington it had been relativity theory. At Caltech it was a new development in mathematical logic. Since the days when he had studied *Principia Mathematica,* he had had a kind of avocational interest in the subject. Now, in 1933, he scheduled a course emphasizing the recent work of the Polish logicians, Jan Lukasiewicz and Alfred Tarski, on many-valued logics.

Because logic was outside Bell's special interests in algebra and arithmetic (although appealing, as he once wrote, to the type of mathematician drawn like himself to the *discrete* rather than the *continuous*), I consult the logician Solomon Feferman about this subject for which Bell exhibited such enthusiasm. Feferman is well qualified. As an undergraduate at Caltech in the mid-1940s he received his introduction to logic in Bell's course and later took his Ph.D. under Tarski.

"Bell's course was not taught in any systematic way," Feferman recalls. "Instead we sampled a lot of sources about a mixed bag of topics. I don't think that Bell had any sense himself about what the core of the subject was. It was only when I got to Berkeley as a graduate student in 1948 and took Tarski's course in Metamathematics that I understood that there was a comprehensive and systematic way of looking at the subject of logic."

"What did Tarski think about many-valued truth by then?"

"I think he found it of technical interest," Feferman says, "but I don't think he ever took it as seriously as Bell did. As it happens, many-valued truth has had no impact on mainstream logic. It just doesn't correspond to any natural form of deductive reasoning. If we don't know that something is certainly true or certainly false, would we assign a truth value of 1/2 to it? That is what is done in the simplest many-valued logic of Lukasiewicz and Tarski, and then one calculates truth-value combinations. A more sensible many-valued logic was to

be introduced in the 1950s by Kleene to describe computational phe-
nomena, using the three values: True, False, and Undefined (or Un-
determined). Also Brouwer's intuitionist logic (really due to Heyting)
has proved to be much more significant than either of these, for it corresponds to methods of reasoning in constructive mathematics."

But Bell was enthralled with the idea of "truth" as something in-
determinate. When Robert Gill of the Williams & Wilkins Company
asked him for another book on science, he offered one tentatively titled *Pilate Silenced*. The thrust of the book was to trace ideas about
mathematical and scientific truth and to argue that there is no "absolute" truth. For his epigraph he took John 18:38—"Pilate saith unto
him, 'What is truth?'"

It appears from this book (which was published as *The Search for
Truth*) that the exuberance Bell had earlier funneled into his scientific
adventure stories was now being funneled into a type of popular science book quite different from *The Queen of the Sciences*. *The Search for
Truth* is a boisterous book. The author is even more present than is
usual in Bell's popular writing. One has the sense that one is with him
in his little "shack" as he writes. He is constantly referring to his own
preferences and aversions, reminiscing about the old Barbary Coast,
describing drunken binges in San Francisco. At one point he gives the
date and time of his writing. At another he jumps up to check the
birthplace of Boole. There are constant interruptions by a character
called "Toby." In fact, most disturbing stylistically is a chapter in which
he brings in his fictional "Toby" to challenge members of the academic
establishment (something that the real Toby was not at all averse to doing). In the final chapter he descends to slapstick as he describes two
of his characters, the Dean and Mrs. Moody, falling into a bathtub in
the heat of their debate over who is to deliver the address on religion:
"As chairwoman, Toby ruled in favor of the Dean, chiefly because he is
a gentleman and Mrs. Moody obviously is not."

In the case of the book, as in the case of Bell himself, opinion is
sharply divided. At the time a *New York Times* reviewer found it "a
brilliant book, and one so much out of the usual in its method and
attitude that it is a bit breathtaking." But today, to a logician like Feferman, it seems "so larded with all sorts of jokey asides and little stories
that you can't believe he's serious. There is very little real history. It
is quite jumpy and is always going off on tangents. Only in the last
few chapters does he try to get more serious and summarize his views.
He lists four great steps in the history of mathematics. These are really

very curious since, while the first three would be generally accepted, most mathematicians would feel that many important steps (although referred to in the text) have been omitted and that his fourth and final step, the discovery by Lukasiewicz and the followup by Lukasiewicz and Tarski of non-Aristotelian logic, is not one that they would subscribe to."

A fan of *The Search for Truth* was Bell's friend the book dealer Jake Zeitlin. Interviewed in 1981 by Alexanderson, he recalled how he had mischievously sold a copy to Remsen Bird, a former minister who was then president of Occidental College.

"Remsen Bird was very much wrought up by that book. In fact, he went to Albert Ruddock, who was the Chairman of the Board of Trustees at Caltech, and Ruddock in turn went to Millikan (who had been the object of ridicule, good-natured ridicule, in Bell's book) and said, 'We simply can't have a man like E. T. Bell here.' It was Bell himself who told me the story. He said that Millikan said to Ruddock, 'Today you don't like E. T. Bell because of his views and we fire him. Next week we fire Paul Epstein because he's a Russian Jew. Then we fire Graham Laing because of his unorthodox views on economics. First thing you know this will no longer be a first-class institution.' Bell was very impressed by that."

Bell signed the contract with Williams & Wilkins for *The Search for Truth* the same day that he signed a contract with that publisher for his science fiction novel *Before the Dawn*. Williams & Wilkins was essentially a publisher of science books—later it moved exclusively into medical texts—but its president, Robert Gill, was so enthusiastic about Bell's novel that he drew it into the fold under the designation *fanta-science* (a name suggested by Bell) and enthusiastically publicized the newly created genre.

In *Before the Dawn*, through the invention of a "televisor," a small group of men are able to view the earth in the Mesozoic and to observe, ultimately, the last days of a small group of dinosaurs. Bell's strong emotional attachment to the book—his favorite, so he wrote in *Twentieth Century Authors*—clearly derived from memories of his boyhood collection of lizards and horned toads and his adolescent visits to the life-size dinosaur models on the grounds of the Crystal Palace Exhibition in Upper Norwood. Although he referred to them in *Twentieth Century Authors* as "those great amoral brutes," he invests the dinosaurs in *Before the Dawn* with human traits and virtues. His hero, Belshazzar, is loyal, courageous, and inventive; and the title of the

ANNOUNCEMENT of the publication of a book of "fantascience"

chapter that describes his end is from the requiem, "Sunset and Evening Star."

I cannot date the writing of *Before the Dawn* but can only say that it is never mentioned in any of Bell's pre-1930 letters to Dutton as being written or as being a plot in mind. In May 1930 Bell wrote to Hughes that he was "about ready, after a rest, to try another story." The device of "the televisor" points to the early 1930s when television broadcasting was an exciting prospect in America—it had already begun in England in 1926—but the scientific spirit of the book and its emotional quality are of an earlier time. In his essay on *Before the Dawn* in Frank N. Magill's *Survey of Science Fiction Literature,* Stableford points out several ways in which the book emerges from the nineteenth century.

"The novel not only is an example of the impact scientific discoveries can have on the popular imagination in general [Stableford is referring here to the discoveries of Cope and Marsh in the American west in the 1890s], but is a testament to several Victorian attitudes in particular, such as an unabashed admiration for power, even when it entails bloodshed; a celebration of competitiveness and ferocity; and a certain occasional reticence—even hypocrisy—about revealing one's emotions or true opinions."

In many ways, in spite of Gill's enthusiasm for "fantascience," *Before the Dawn* is the least "scientific" of Bell's science fiction novels. The dinosaurs observed are never identified by species, and not one of the four characters privileged to observe life in the Mesozoic is a scientist capable of appreciating what he is seeing. ("The only scientist among us" is a social anthropologist specializing in the Mayas.) Although the four call on museums for sample fossils, it never occurs to them to call in a paleontologist; odd, since one of Bell's good friends at Caltech was the paleontologist Chester Stock. Instead they interpret the prehistoric events they are witnessing in terms of their own emotions and motivations and their own conceptions of good and evil. Reading the account of the final battle between Belshazzar and Satan, one is tempted to wonder if Bell is not telling again a story he made up when he was playing with lizards under the fruit trees of the Santa Clara Valley.

In spite of its defects, *Before the Dawn* contains so many scenes of haunting grandeur that it has continued to grip readers. In his chapter titled "The Vision of Science" in *The Strength to Dream: Literature and the Imagination,* Colin Wilson singles it out for extensive discussion and offers it as a classic of "the purest type of science fiction"—that which "attempts to evoke the same type of wonder that watching a chemical experiment produces in an eleven-year-old."

(Or as John R. Pierce, father of Echo and Telstar and formerly the head engineer at Bell Labs, remarks to me: "Science fiction serves the same purpose for science that miracles serve for religion.")

Bell also stated in *Twentieth Century Authors* that *Before the Dawn* was the only one of his novels that he published under his own name. This statement conveys a slightly inaccurate impression. The title page carries the inscription "By John Taine (Eric Temple Bell, Professor of Mathematics, California Institute of Technology)," but by then the fact that John Taine was E. T. Bell was common knowledge.

At the time of its publication, *Before the Dawn* was reviewed in the mainstream press as well as in science fiction magazines. Elmer Davis in the *Saturday Review of Literature* compared it to H. G. Wells's early fantasies "and as good as all but the very best of them." An unnamed reviewer for *Wonder Stories,* while complaining that John Taine took almost half the book "to get down to brass tacks," found it "filled with breathless action and vivid adventure." C. A. Brandt, reviewing it in *Amazing Stories* with a book "said to be the autobiography of a drug addict," treated both as "something other" than science fiction.

"[But] you can always take for granted," he assured his readers, "that anything John Taine writes is not only interesting, but it is always based upon a unique and absolutely new Idea [sic]. And, whatever the idea, and no matter how farfetched it may seem at first, it is always presented and worked out in a perfectly plausible manner. He writes convincingly at all times, and never uses any 'literary crutches,' or leans upon somebody else's ideas. He is always 'John Taine' and 'John Taine' only."

This last sentence suggests the following question, which I propound to various members of the science fiction community:

If you picked up a novel by John Taine, not knowing the author, would you mistake it for a novel of any other writer?

Answers range from "Absolutely not!" from Julius Schwartz, Bell's agent for his science fiction after 1935, to less unequivocal responses from other writers, most of whom are understandably a little vague on the subject, since they have not read John Taine for many years. The most interesting answer comes from Stableford.

"I am unsure of the relevance of your (rather peculiar) question," he writes, "—unless perhaps you have found evidence that Bell did not write all of the 'John Taine' novels . . . ."

The "heterogeneity" of the work, he explains, makes my question difficult to answer. Although Campbell in his discussion of John Taine describes him as strictly "a formula writer," Stableford finds novels like *Green Fire* and *Quayle's Invention* "similar to other items in the tradition of near-future thrillers . . . , but the mutational romances [these would include such novels as *The Iron Star* and *Seeds of Life*] are much more distinctive and much more significant—they helped considerably to extend the imaginative horizons of pulp sf." He finds *The Time Stream* still unique.

Stableford's response is stimulating. How much more difficult would it be, I wonder, if I ask a similar question and frame it more generally:

If you picked up a piece of writing by E. T. Bell, not knowing the author, would you mistake it for the writing of anyone else?

Taking each piece individually might provide some difficulties, but I cannot imagine that anyone other than E. T. Bell could have published in one year (1934) three pieces so completely different from one another as *The Search for Truth, Before the Dawn,* and two mathematical papers that appeared that same year: "Exponential numbers" and "Exponential polynomials."

These last papers fall in the third group of Bell's mathematical interests: the studies of Special Numerical Sequences and Functions. Works in this group are for the most part concerned with questions about partitions—very roughly, expressing numbers as sums of other numbers. A variant on the question of partitions of numbers is the question of partitioning sets: how many ways can a finite set be split into subsets none of which overlap and which together contain all the members of the original set? For a number n equal to or greater than one, the number of different partitions of a set of n members is known today as "a Bell number."* Bell's interest in such numbers had been aroused when, in some calculations he had been asked to check, he had found a discrepancy and had traced it back to the mathematical handbook used, in which the MacLaurin expansion had been given incorrectly. It happens that the coefficients for x in that expansion are,

---

* Bell did not discover these numbers, which have been known since ancient times, but he was the first mathematician to treat them seriously and reveal their mathematical importance. The reader interested in a popular treatment of the subject that will allow him to play with Bell numbers, and perhaps make some discoveries about them, is referred to Martin Gardner's *Scientific American* column on "The Tinkly Temple Bells," which is reprinted in his *Fractal Music, Hypercards and More.*

in fact, the number of different partitions of a set S_n; in short, the Bell numbers. The error that Bell had found in the calculations suggested to him "the desirability of having some readily applicable numerical check on the tedious algebra involved in expanding functions of the type . . . , and this in turn led to the definition of *exponential integers* and the investigation of their simpler mathematical properties . . . ." The exponential integers of the first paper then suggested extensive generalizations to *exponential polynomials,* and these resulted in the second paper on what are known today as Bell polynomials.

So mathematics grows.

At the time Bell published these papers the invention of the electronic high-speed computer was still more than a decade away, as was the resulting development of increased interest in discrete mathematics. The importance of this work of Bell's for contemporary combinatorialists can be gauged by the fact that "Exponential polynomials" is, thirty years after his death, the most frequently cited of his papers.

By this time in his career Bell was doing much more writing not directly connected with mathematical research than he had done in the past. Angus Taylor tells me that once, as a young instructor, he complained to Bell that his teaching, "which I enjoyed and to which I gave high priority, was leaving me with much less time for study and research than I would have liked to have. I was shocked when he said, 'Well, Taylor, you'll just have to neglect your teaching a bit.'"

In the year following the appearance of *The Search for Truth* and *Before the Dawn,* Bell published eight mathematical papers, several general articles, and a number of book reviews, both in mathematical journals and in the local newspaper. In the latter, on occasion, he enjoyed reviewing a book by E. T. Bell under the byline of John Taine. One reader did not get the joke and complained that a science fiction writer was hardly qualified to review a book by Dr. Bell.

That same year (1935) Bell for the first time employed an agent for his novels. This was Schwartz, who with a partner had recently founded the first literary agency to specialize in science and fantasy fiction. Bell gave Schwartz a novel written apparently after the inauguration of Franklin Roosevelt in 1933, since it features an American president who, like the polio-stricken Roosevelt, spends a great deal of time in the swimming pool. Schwartz promptly sold *Twelve Eighty-Seven* to *Astounding Stories* for $540 (a cent a word), and the first of five installments appeared as the lead feature ("A Thought Variant Epic") in the May 1935 issue.

The plot of this novel may be one Bell sketched to Dutton at the end of 1929: a kind of sequel to *Green Fire* "with an actual 'war of the worlds' based on strictly modern scientific principles." Taking off from the contemporary "dust bowl" of the American southwest, he postulates a "miracle" dust invented by the Japanese that, when mixed with the depleted American soil so improves fertility that it puts fertilizer companies out of business. Ultimately though, unknown to the Americans, it is designed to render the treated soil sterile. Characters are Jay Jarvis, a young chemist who has "extended" the periodic table (the number of the dust on the scale is 1287); his older Japanese friend Tori, who turns out to be involved in the Japanese plot to use the miracle dust to control the world; and Tori's beautiful sister, Nara, a pacifist and an internationalist. Catastrophe is averted only when Nara and her supporters sink the ships carrying the dust, "deadly to all life." How to prevent dominance by a country making such a discovery in the future? Jay's solution, being broadcast at the end of the book, is the scientist's answer: tell everybody everything.

Since Arthur C. Clarke in *Astounding Days* dismisses *Twelve Eighty-Seven* as "inferior," I look to see how readers reacted. "Please, whatever you do, don't stretch it out to six parts. Have pity on us readers." "It certainly adds to the prestige of the magazine to have such an author between the covers." But enthusiasm was not unanimous. "It is really not a science fiction story. Taine had better mix in some future dates, some rocket ships, queer inhabitants from other planets, and ray guns if he wants to win my respect as a science fiction writer."

Bell, however, was never much interested in writing about mechanical things, and by this time he had apparently lost interest in writing scientific adventure stories, or no longer felt the drive or the need that he had felt in the past. In February 1935, writing to Hughes, he mentioned that he was "in the middle of a new book." Six months later, in July, he signed a contract with Simon and Schuster for a book titled *The Lives of Mathematicians*, which he promised to deliver November 1.

It would be published as *Men of Mathematics*.

DESERT LANDSCAPES in vivid colors were favorite subjects

274

The Human Side of Mathematics

It has been said, possibly by Bell himself, that the human side of mathematics is mathematicians. The statement could have served as an epigraph for *Men of Mathematics*, which was proceeding through the various stages of publication between 1935 and 1937. Thus it seems reasonable at this point to try to obtain—from people who knew Bell during the period—a glimpse of the man at home with his wife, his son, his friends and colleagues, as well as with students.

Of the latter, Hubert Arnold, although a student of Michal's, was closer to Bell than most of the young people. The fact that his "Uncle Rob" was R. E. Moritz, the "Pterodactyl" of Bell's days at Washington, provided Arnold with his entrée to the Bell household. He confesses, however, that what really endeared him to Bell was his saying that he wanted some teaching experience—he was "tired of sucking at the journal tits."

Young Arnold was something of a kindred spirit. At the University of Nebraska he had been a student of M. A. Basoco, a doctoral student of Bell's for whom Bell had had a special fondness—the only mathematician with whom he ever published a joint paper. Basoco had urged Arnold to go to Caltech for his graduate work, but that young man had wanted first to spend a year in Germany. It was 1933, Hitler had just come to power, and he was advised by knowledgeable people "to wait until things blew over." Instead he went to Paris for a year and then, in the fall of 1934, to Caltech. Later, after abandoning a Princeton fellowship with Lefschetz and spending four years in the Navy, he

HUBERT ARNOLD
a member of the family

joined Martha Graham's dance company—"rising to understudy" and
becoming well acquainted with Miss Graham since he was older than
other members of her company. Ultimately, of course, he returned to
mathematics.

Young Arnold was in and out of the Bell house all through his
years at Caltech. In a very short time it was "Romps" and "Toby" and
"Brother Arnold" ("because Romps and Toby thought I looked like a
preacher"). Because of his intimacy with the Bells and the five-year
span of his time in Pasadena, I hope that I will learn from Arnold how
long Bell's classic *Men of Mathematics* was in the planning. But Arnold
does not remember Bell's ever talking about what he was working on,
although he remembers very well how Bell worked.

"He worked all the time. After dinner he would disappear and
work straight through the night. I don't know when he ever slept."

Oddly, even in twenty-three letters (from 1912 to 1936) to the historian of mathematics D. E. Smith, there is no reference by Bell either
to the conception or the composition of a book about the lives of mathematicians until *Men of Mathematics* was on the verge of publication.

From Bell's letters at this time, it is clear that he was not at all happy
with Simon and Schuster as publishers. He complained that it had
taken them almost as long to get the book printed as it had taken him
to write it. Then, wanting something to tie in with their *Men of Art*,
they had put a title on it that he did not like. Worst of all, they had
insisted that he cut the manuscript by 125,000 words.

"[They] said they would have to charge $10 for it unless it was cut; so I took out over a third," he complained. "They plan now to get it out for $5—too much, in my opinion."

Some idea of the extent of the cut Bell made can be seen on page 237 of the published book. There he has just described how Dirichlet carried his copy of the *Disquisitiones Arithmeticæ* on all his travels, struggling with some tough paragraph before he went to bed and sleeping with the treasured volume under his pillow. "Dirichlet did much more in mathematics than his amplification of the *Disquisitiones*," Bell goes on to say, "but we shall not have space to discuss his life. Neither shall we have space (unfortunately) for Eisenstein, one of the brilliant young men of the early nineteenth century who died before their time . . . ." A footnote on the same page refers to Legendre and the fact that "considerations of space preclude an account of his life."

Even while he was reading proofs and obtaining photographs for *Men of Mathematics,* Bell continued to write. It was as if he had to write, according to those who knew him—the words just seemed to flow from his head to his hand and onto the page. Between November 1, 1935, when he submitted the manuscript of *Men of Mathematics,* and March 2, 1937, when he received his first printed copy, he had written and had in print *The Handmaiden of the Sciences,* a sequel to *Queen of the Sciences,* the little book he had written for the *Century of Progress Series.*

In the middle 1930s money for education was very tight. Mathematics, along with Latin and even modern languages, was under attack as an educational "frill." The situation was disturbing to all mathematicians, but it had brought Bell to the barricades, not only to defend mathematics but also to castigate his fellow mathematicians for their failure to do so.

"[I]t must now be obvious, even to a blind imbecile, that American mathematics and mathematicians are beginning to get their due share of those withering criticisms, motivated by a drastic revaluation of our ideals and institutions, which are only the first, mild zephyrs of the storm that is about to overwhelm us all . . . . In the coming tempest only those things will be left standing that have something of demonstrable social importance to stand on. Mathematics, as we mathematicians believe, has so much of enduring worth to offer humanity on all sides from the severely practical to the ethereally cultural or spiritual that we feel secure—until we stop to think."

In his opinion, mathematics had not presented its case in a way "that the men and women who pay the bills which make mathematics possible can see clearly what they are asked to pay for." This had to be done "and immediately, if mathematics is to survive in America." His own contribution was the book *The Handmaiden of the Sciences*.

In *Queen of the Sciences* he had sketched the nature of mathematics itself, especially as it had developed in the past century. Now he wanted to tell what he called "the other side of the story." The new book was to be "the narrative of how mathematics has served the sciences and, through the sciences, mankind."

Unlike *The Search for Truth*, which Bell himself had dismissed as "a semi-serious fling," *The Handmaiden of the Sciences* was a serious attempt to convey to the public the practical value of mathematics—but not of course serious to the point of dullness. The *New York Times* found it a book "[that] both professional and amateur devotees of the oldest science . . . will esteem highly." Mathematical reviewers agreed that at a time when mathematics was "under attack," the little volume would be "a useful campaign book."

As he had done in *The Search for Truth* and would do in *Men of Mathematics*, Bell acknowledged in his preface "the invaluable help of Dr. Edwin P. Hubble and his wife, Grace." Edwin Hubble died in 1953 and his wife in 1970, so I am not able to talk to them about Bell. At the time of Bell's death, however, Grace Hubble wrote to Taine and Janet about how she and her husband had felt about "Romps."

"With all the wit and brilliance of his talk and its wide ranging, our friendship was unique, and in a way absurd, for it took us from the stupidities of common day into a fantastic freedom of imagination, like games children play when grown-ups are more than usually boring. Nobody else could play this game for either of us again."

As was the case with most couples in his circle, Bell was closer to the wife than to the husband. Grace Hubble, like Luddye Michal, was one of that quite large group of women friends in whose company he delighted, and who delighted in his company. What was the source of his attraction for them? Certainly he was good-looking. He could be very charming. (But also crude in a kind of adolescent way.) He respected their intelligence and treated them as friends. The adjective I heard most often applied to him by women was *interesting*. Bell himself may have offered his own explanation of his attraction for them in *The Gold Tooth* when, analyzing the reason for his heroine's infatuation with her cold Japanese lover, he suggested that "what attracts young girls to the haughty Byrons and the cold Okadas of society is not intriguing rakishness or fascinating aloofness, because these qualities are common enough . . . but something much rarer—sheer intellect."

Grace Hubble's journals—like Bell's letters to Hughes—enable me to date certain anecdotes of these years. One of these relates to Bell's contact with Aldous Huxley and other writers of the English colony then on the fringes of Hollywood. In October 1935 she mentions "Eric and Toby's dinner for Aldous and Maria." She makes no further comment; however, Jake Zeitlin, interviewed in 1981, described "a lunch" with Huxley and his wife at the Bells' house in more detail.

"Bell very quickly rebuffed Huxley. He didn't like Huxley's trafficking with mysticism. Also Huxley was very much involved as a friend with Gerald Heard, who was a very brilliant Englishman but given to homosexual companions. Bell really had no use for that sort of thing. In some ways he was a man who was at home with great ideas, and in other ways he was a very provincial man."

(In this connection the space scientist Al Hibbs remembers Bell telling a class in 1945, "The Greeks left two great concepts for western civilization: homosexuality and extrapolation; of the two the latter has done by far the more damage.")

It was during the time when Arnold was so close to the Bells that H. P. Robertson came to spend a year at Caltech on a sabbatical from Princeton. This was the first extended period during which Bell and "Hot Dog" had been together since their days as teacher and student at the University of Washington. I am curious about the relationship between them after the break in 1929 when a deeply hurt Robertson resigned the position at Caltech he had so eagerly accepted earlier. I ask Arnold his impression of their attitude toward each other.

"Oh they were great friends!" he says. "Both Romps and Toby were very fond of Robertson." He smiles at the memory of the younger man. "He was such a fellow! Really big, over six feet, and heavy. Jovial. Always joking. Smoking big cigars and drinking, too, I guess. He always seemed to me more like a businessman, a Chamber of Commerce type, than a scientist."

Arnold finds it hard to believe, from his own observation, that there was ever the slightest cloud over the relationship between Bell and Robertson. Others, who saw the relationship more from Robertson's side than from Bell's, have a different view.

"Bob had very deep feelings in regard to Bell, and they oscillated," I am told by Abraham Taub, who was close to Robertson during his own graduate days at Princeton in the early 1930s. "In a way Bob was *beholden* to Bell. [This was a word also used by Jesse Greenstein, a later student of Robertson's, to explain the relationship.] I don't think Bob liked that. There was a love-hate relationship there. I think one of the reasons was that Bob had been very hopeful that on his return from Europe in 1929 he would go back to Caltech, and he regarded the fact that he didn't as a betrayal on Bell's part. It was something very deep in Bob's psyche. But he always wanted to go back to Caltech. And later, when he had an opportunity, he didn't think twice."

During the year of Robertson's sabbatical, he and his wife, Angela, lived at the Athenaeum, their children remaining in the East. They frequently saw Bell and Toby, often dining with them. The story that follows pertains to one such occasion. It is told to me by Robertson's daughter, Marietta Fay.

"I, of course, was not there," Mrs. Fay explains, "but I have heard my mother tell this story so many times that I know I am repeating her words almost verbatim."

According to Mrs. Fay, the year her parents spent at Caltech, Taine was attending UCLA. Although he had originally intended to major in mathematics there, he had very soon found that he thoroughly disliked the subject. When, having solved a problem in number theory on his own, he was accused by his professor of having received help from his father, he made up his mind to change majors. His father, however, was insistent that he continue with mathematics. In fact, he was so adamant on the subject—and probably quite cruelly sarcastic, as he could be—that the young man was seriously upset. His mother, worried about his emotional state, decided to take him to the desert for a few days to get him away from his father. She called Angela Robertson, explaining the situation, and asked her, since Bell would be alone in the evening, if the Robertsons would invite him to have dinner with them at the Athenaeum.

"Well, all through dinner, Bell just went on and on about Taine," Mrs. Fay continues. "How stupid he was and so forth—it was awful—and finally my mother turned on him and said—these were her exact words—'Dog Romps, if I were your son, I would just go down to San Pedro and get on a freighter and never come back!'

"Bell looked at her for a minute with that icy stare he could get on, thin lips in a line, and then he stood up and, without a word, stalked out of the dining room.

"Well—when Toby got back, my mother went over to the house to explain what had happened. Toby was washing dishes, a cigarette hanging out of the corner of her mouth the way one always was, and she practically let it drop into the dishwater when she heard what my mother had said to Bell that evening.

"'You said that to Romps!' she exclaimed. 'How did you ever happen to say that? Because that's exactly what Romps did—he took a ship and came to America to get away from his father!'"

So there again is the story that Toby had told Taine about why his father had left England, a story that Taine told me on the occasion of our first meeting. At the time I thought that Taine might have misunderstood Toby, or that she might have misunderstood her husband— that it might, in fact, have been Bell's father, James Bell Jr., about whom she was talking. For he had left home and he had come to America, quite possibly to escape the domination of that powerful figure at Penybryn, James Bell Sr. But no, from Mrs. Fay's account it is clear that Taine was correct. This was the story his mother had told him about his father's coming to America, because this was the story she thought was true.

Later I recount Mrs. Fay's story to Taine. It appears, I say, to support what I have begun to suspect—that even Bell's wife never heard the story of his boyhood in San Jose.

Taine agrees.

"If my father had told her," he says, "I know that my mother would have told me."*

In connection with Mrs. Fay's account, I ask Taine about what eventually happened in regard to changing his major.

"In the summer of 1936 I took a course in biology and I really liked it, so I switched to biology," he replies. "My father was angry, partly because it would require my going to UCLA for another year, but he was sure that biology was the coming thing. I think he got that idea from Thomas Hunt Morgan, who lived across the street from us."

A classmate in Taine's biology class that summer was a young woman named Virginia Lee (now Cornwall). From her I get another glimpse of the Bells' domestic life at the time of *Men of Mathematics*.

"Sometimes Taine would take me back to Pasadena for the weekend. He had an orange Ford roadster, and on the hood his mother had painted a dragon. It was pretty spectacular. The first time, before we

---

* In 1955, after Toby's death, Bell wrote in *Twentieth Century Authors* that he left England "to escape being shoved into Woolwich [the training academy for officers in the British Navy] or the India Civil Service."

MADELEINE OLIVER DMYTRYK, a friend from the movie colony

went, he explained to me that his father 'came on strong,' but that I shouldn't be offended—his bark was worse than his bite.

"Taine called his parents Romps and Toby, and they wanted me to call them that. I could say 'Toby,' but I couldn't say 'Romps.' I always said 'Dr. Bell.' Finally he said disgustedly, 'You might as well call me sir,' so I called him 'Sir Romps.' He had nicknamed me 'Bubbles' because I was an effervescent, Pollyanna-type at that time. After I started calling him 'Sir Romps,' he called me 'Lady Bubbles.'

"Sometimes when I came he would have been drinking, or he would do something outrageous—like farting and looking at me to see how I reacted—Taine would just look disgusted, and Toby would say, 'Hush, Romps.' But he was a wonderful man—so interesting. It was hard though for Taine being his son; he had this feeling that he could never come up to him. But Taine and his mother had a wonderful relationship. They would dance together around the kitchen, and juggle things, and joke and tease."

Even today the memory of the Bell home is vivid in her mind: the jungle-like garden, the big bromeliad in the oriental pot on the front porch, the dinner plates bought in Chinatown, the many pictures—Toby's small and inclined to be pastel, his large and vivid, often desert scenes. He seemed to have a liking for the desert, she says—and lizards.

"Taine and I and his parents became a regular foursome that summer. We would go to unusual restaurants and to plays at the Community Playhouse and to concerts at the Hollywood Bowl. It was wonderful. They didn't mind driving. They thought nothing of coming to Manhattan Beach, which is quite a distance from Pasadena, to pick me up or take me home. I loved them all!"

Another young woman whom Taine introduced into the family group was Madeleine Oliver, an actress whom he had tried to date before finding out that she was married. There are two glossy photos of her among Bell's memorabilia, one inscribed to him and the other to Toby ("On account of I wish I could be more like you").

There were of course a number of people who did not like the Bells, and who especially did not like Bell. Most of these either deny to me that they have any recollections of him or are just very, very polite in talking about him. An exception is Clifford Truesdell of Johns Hopkins University who, responding to my request in the *Mathematical Intelligencer*, volunteers recollections, "mostly unflattering," he warns. Although Truesdell's time at Caltech, where he took his master's degree under Bateman, was somewhat later than the period I am writing about, I am including his comments here for balance.

"In and around Los Angeles the California Institute of Technology was thought to be the world's center of science," he writes me. "Small wonder, for a persistent local rumor kept Einstein located there, though in fact he was long gone. Millikan, too, and other resident Nobel Laureates were often mentioned . . . , and E. T. Bell, especially after his *Men of Mathematics* had appeared on a list of best sellers, shone not much less in the public eye.

"Of course a boy born and bred an Angeleno, resident but a brisk walk from the Apollonian campus, expected to find there the world's supreme mathematician. In my fourth year I entered Bell's classroom expecting great revelations. Each student was armed with a textbook, the material in which Bell assigned piece by piece to them for presentation at the blackboard. Bell was bored, and soon so were we. If a student got stuck, Bell could not help. Instead, he got angry at the book. He would then slam his pince-nez on the table in front of the blackboard, and many a skid it survived. Perhaps he had practiced. At least he was democratic, taking his turn at the board, and never able to explain anything that stopped his students."

CLIFFORD TRUESDELL
unfavorable recollections

Truesdell deplores Bell's loud clothes, "fit for a salesman of used cars," his positive opinions ("Once at dinner at my house, seeing on a shelf some books by ancient Greek authors, he said, 'I hate the Greeks.'"), his "scurrilous style" in *Men of Mathematics*, which was "a kind of bible" among the mathematics and physics students. He quotes with approval a colleague's statement that Bell wrote every great mathematician's life as if he were writing a script for a movie starring Paul Muni. He also points out that in the entry on Bell in the *Dictionary of Scientific Biography*, written by the mathematical historian Kenneth May, "not a single contribution to mathematics itself is credited to Bell," only *Men of Mathematics* and *The Development of Mathematics*. (May used the latter book as a text in his course on the history of mathematics.) In fairness to Bell, it should be said that May's account is more "a life" than a scientific biography. Other than the mention of the Bôcher Prize, there is only the following sentence about Bell's mathematical work:

"At the University of Washington from 1912, Bell published a number of significant contributions to numerical functions, analytic number theory, multiply periodic functions, and Diophantine analysis."

This is a surprisingly inaccurate statement for a historian like May to make since, as an example, Bell did not even work in analytic number theory. May also inaccurately places all Bell's mathematical contributions at Washington and implies that nothing significant was produced at Caltech, which is where Bell did all his important work in multiplicative diophantine analysis. In addition, the writeup ignores how significant Bell's work on numerical sequences and generating functions has been for combinatorialists.

By the time of which I am now writing (1935–1937), Bell had abandoned the regimen that his doctor had imposed following the heart attack at Christmas 1932. Coffee by the jugful was again carried out to the shack every night. His wallet was stuffed with cigarettes butts.

"I often saw him take one out, light it, and get the last few puffs," Arnold recalls. "He was very frugal."

He had also begun to drink again. The Eighteenth Amendment (Prohibition) had been repealed in 1933, and he could now consume hard liquor instead of the beer he and Taine had been making. When he drank, so Toby complained to Arnold, he became "nasty." In fact, during this period the situation in regard to his drinking became so bad that she went over to UCLA to discuss with Taine what she should do about it. At one point she even threatened to leave, and did leave for a week or two. In the end, however, she came back and decided to let him "hang one on" every so often.

"Otherwise, except for her saying that he was nasty when he drank, I never knew of any unpleasantness between them," Arnold says. "She took care of his clerical work, appointments, letters, telegrams, arrangements, and so forth. There was a very clear agreement as to who did what in that household. But they had a lot of common interests. They were really just like one person."

In the course of our conversation he repeats the last sentence several times.

Since Toby was so important in her husband's life, I ask various people to describe her at this time, and also to describe the relationship between the two of them.

The year *Men of Mathematics* appeared, she was sixty years old. Physically she was of course still short but no longer "tiny." She had become a little plump, and the hair that Bell described to Arnold as having been "the color of the inside of a copper pot" when he met her was now completely gray. She hennaed it, a bit carelessly, the gray sometimes showing at the roots, and always wore bright red lipstick, which often looked as if she had not glanced at the mirror when she put it on. There were members of faculty society who felt that "Mrs. Bell did not have the best advice on her clothes" and was "not blessed with the talent for dressing herself in a way that made a pretty picture."

Lehmer and his wife, Emma, found her almost as unpredictable as Bell, with whom they were always inclined to be "cautious" over the many years of their acquaintance.

She was clearly an intelligent, vivacious, and loving woman who shared her husband's interests in poetry and the arts and was indispensable to the functioning of his household; yet there were people who wondered why a man so obviously attractive to women as Bell—and so obviously attracted to pretty young women—had chosen to marry her. To Dilworth, however, it seems clear that what attracted "E. T." to Toby was that "she could give back as good as she got."

"Toby was just something else again," Arnold agrees. "She would say or do or write anything."

He recalls her describing someone as "looking as if he smelled a stink," then pausing a moment and adding, "Come to think of it, Romps looks rather like that." On another occasion, at a lecture by a pompous "moon faced" education professor from UCLA, she stood up while he was speaking and announced that she herself had attended

Columbia's famous teachers' college that he was praising so highly and she could tell him that it was "rotten." After Arnold left Caltech she would write to him that her son "is happier now that he has found a place to cool his testicles."

From Arnold and from other people I talk to, there is never the slightest indication that any of Bell's many friendships with women other than his wife were anything but platonic. Still I am curious about how Toby felt about such friendships.

"I think she was proud that it was 'her man' to whom other women were so attracted!" Arnold responds with confidence. "I don't think there was ever any trouble between them on that."

In the spring of 1937 *Men of Mathematics* finally appeared. Although Bell had not been happy with Simon and Schuster as publishers and was disappointed in the fact that the book was not so complete as he had planned it to be, it is clear that he considered it a book into which he had put his heart. It was the only one of the many books he had published up to that time that he dedicated to someone.

"Toby, as before, has contributed much. In acknowledgment for what she has given, I have dedicated the book to her (if she will have it)—it is as much hers as mine."

In the first copy he received from the publisher he added a few more words around "To Toby," inserting "Dog" before "Toby," so that the dedication then read:

> To Dog Toby
> from
> The Author Dog Romps
> 2 March 1937
> Pasadena, California

There are similar inscriptions to Toby in all the books published by Bell up to that time. The first copy he received had always gone to her, but it puzzles me that never in any one of them—and I have seen them all—did he write "With love"—it was always some such phrase as "With the Author's respects" or "With the best regards of the Author."

I ask Arnold what he thinks was the reason.

"Bell was a very proper man," he replies. "He didn't like overt expressions of emotion."

"The love was there," according to Taine, "but he could not put it in words."

MORGAN WARD: "It would be a great mistake to let Ward go."

The Archives, California Institute of Technology

Of a Man and Mathematics

Men of Mathematics was to become an instant classic and, if not the most respected of Bell's books, the most popular. Mathematicians read it and enjoyed it—one, H. F. Bohnenblust, read it aloud to his nonmathematical bride on their honeymoon. Students read it and enjoyed it. Many got from it their first glimpse of mathematics as something different from and immeasurably more exciting than "math" and, as a great number later told their professors, decided as a result to become mathematicians. Members of that vague, immeasurable group of educated laymen (the existence of which is a tenet of faith with scientific popularizers) read the book with pleasure, although never so many as Bell hoped—he once defined a scientific popularization as a book read and enjoyed by those who don't need it and not read by those who do.

The book was favorably reviewed in the contemporary press. In fact, the *Book Review Digest* of 1937 mentioned only one critical comment. Bertrand Russell in the *New Statesmen and Nation*, while generally favorable, noted "an occasional suggestion that the present age is wiser than its predecessors, and a slight tendency to regard matters which are still *sub judice* as decided."

A review that must have meant a lot to Bell was that by his Columbia professor, David Eugene Smith, the historian of mathematics. Titled "On a Remarkable Contribution to the History of Mathematics," it opened the spring issue of *Scripta Mathematica.*

"There has recently appeared in this country a new type of history of mathematics . . . ," Smith wrote. The author of this remarkable

H. F. BOHNENBLUST
visitor, later colleague

Office of Public Relations, Caltech

contribution was "a poet who is known as a scientist, and a scientist who is known as a poet." Pondering the question "how a man who is a professor of mathematics at the California Institute of Technology can find the time and be possessed of the literary ability to write such a book," Smith concluded that he must be "so versatile in the affairs of the spirit that without too much exaggeration he may be looked upon as resembling in certain respects Blaise Pascal, minus the latter's invariable purity of style and his unfortunate liaison with theology." He then drew the comparison between Bell's lives and Plutarch's.

The comparison with Plutarch was apt. Bell's biographical accounts were often largely anecdotal, and his successors in the history of mathematics have exposed factual errors and exaggerations seemingly without end. The most glaring is his romanticized treatment of the life of Evariste Galois. Bell's chapter on the founder of group theory, who died at twenty following a duel, did not merely perpetuate a legend. It has since been shown up by the researches of Tony Rothman as intentionally inaccurate and even false in many particulars.

In *Men of Mathematics*, under "Acknowledgments," Bell specifically cited as his source for the life of Galois "the classic account by P. Dupuy . . . and the edited notes by Jules Tannery." These provided the accepted version, and Bell was not alone in relying upon them. But he did more than rely. In "Genius and Biographers: The Fictionalization of Evariste Galois," published in the *American Mathematical Monthly* in February

1982, Rothman showed that Bell very selectively—which is the kindest way one can put it—edited the work of Dupuy and Tannery to support his version of Galois's life as a conflict between "Genius and Stupidity" where genius lost out. This version permitted Bell to indulge in a favorite sport of attacking the governmental and educational bureaucracy as well as the mathematical establishment; to perpetuate several baseless myths about the political and romantic events surrounding the duel; and to write a final paragraph that, as Rothman states, "is likely the very paragraph which has given the greatest impetus to the Galois legend"—

> All night long he had spent the fleeting hours feverishly dashing off his scientific last will and testament, writing against time to glean a few of the great things in his teeming mind before the death he saw could overtake him. Time after time he broke off to scribble in the margin "I have not time; I have not time," and passed on to the next frantically scrawled outline.

While Bell did not actually write in this paragraph that Galois created group theory in its entirety during "those last desperate hours before the dawn" (since he could not ignore the fact that earlier in the chapter he had derided members of the French academy for not appreciating and even carelessly losing some of Galois's revolutionary work), he implied as much. Galois did write a letter to a friend the night before the duel. He did include some mathematics in it. He did "scrawl" beside a theorem, as Rothman points out, "There are a few things left to be completed in this proof. I have not the time," but he also appended to that last statement "(Author's note)."

The above is only one instance where Rothman exposes Bell as omitting material in Dupuy and Tannery, rearranging the chronology of events, and misinterpreting statements, all to support his version of Galois as a passive victim. Bell is not the only writer on Galois to be exposed by Rothman. Under the heading "Harsher Words" he explains the purpose of his article, which "has been to show that something is curiously out of sync. Two highly respected physicists (Fred Hoyle and Leopold Infeld) and an equally well-known mathematician, members of the professions which most loudly proclaim their devotion to Truth, have invented history."*

---

* *Ten Faces of the Universe* (Hoyle) and *Whom the Gods Love* (Infeld).

For the entire account of Rothman's researches, the reader is referred to the *Monthly* article and to the amplified version in Rothman's collection of essays, *Science a la Mode: Physical Fashions and Fictions.*

But to return to the contemporary reaction to *Men of Mathematics*— the chapter on Galois was the one that Smith singled out for praise.

"Here was Galois, a man of whom few readers excepting mathematicians ever heard, with only a single major discovery or invention to his mathematical credit, and this in a branch of analysis that is Greek to the average well-educated man or woman. In spite of this handicap Professor Bell leaves the reader feeling that he knows a good deal about Galois and his great achievement . . . ."

Smith praised Bell's method in all his chapters.

"As to precise historical information he states that his readers must not depend upon the accuracy of his dates, or even upon the incidents mentioned—a misfortune, but a frank confession. He does not seek to document his statements, and thus he avoids the style of the dry-as-dust historian . . . . The pages and the binding will soon decay, but not the impressions which the contents will leave upon its hundreds of readers."

Smith's last sentence has turned out to be quite accurate. In spite of Rothman's revelations and the objections of other historians of science, particularly to the treatment of Laplace and to an offensive remark about mathematicians of the Jewish "race" as being especially prone to controversy, the book has remained highly readable for more than fifty years.

Men of Mathematics was to play still another role in mathematics. In a letter to Bell dated March 6, 1937, the distinguished lawyer and bibliophile William Marshall Bullitt wrote of having gone into Simon and Schuster's offices to obtain a copy, "which they told me was the *first copy* that had been sold." Subsequently Bullitt presented copies to the presidents of the most prominent insurance companies in the country (he was an authority on insurance law and actuarial mathematics) as well as to anyone else he thought might conceivably have an interest in it. His enthusiasm seemed to have no bounds. Three months after the book appeared, he wrote to Bell, "It has inspired me to endeavor to collect the original First Edition of the single greatest work of each of the twenty-five greatest mathematicians."

In this sentence is the origin of what has been called "one of the world's most extraordinary collections of first editions of important mathematical works," the William Marshall Bullitt Mathematical Collection at the University of Louisville. The story of the collection, and the role that Bell and other mathematicians played in it, has been well told by Richard M. Davitt in the *Mathematical Intelligencer*. There is no need to repeat it here except to mention that Bullitt (almost eighty by then) wrote in 1951 to Bell that he, Bell, was "not the Godfather but the actual Father" of the collection.

"I would like to get out a little printed catalogue . . . ," he wrote, "and if I can manage to get it out, it will be dedicated on the Title Page to you."

Bell quite naturally enjoyed the fame and attention that a "best seller" brought. Robertson, running into him in New York, commented to a friend that he seemed to have "calmed down" a little. But he did not rest long on his laurels. Three months after *Men of Mathematics* appeared, he began another book, which he titled *Man and His Lifebelts* and completed it in a month.

Man and His Lifebelts and *The Handmaiden of the Sciences* (which had appeared just six months earlier) represent the two sides of Bell's popular scientific writing. *Handmaiden* is a serious if unconventional attempt to explain the applications of mathematics to young people and to that ever elusive "educated layman." In it a reader will find Bell at his best: a professional mathematician who can write casually about mathematics just because he is a professional. In *Man and His Lifebelts*, on the other hand, Bell is at his most eccentric. As in *The Search for Truth*, he seems to be writing simply for the fun of it, or perhaps one should say for the hell of it, never resisting an impulse to be outrageous in language or ideas. *Lifebelts* is a book to be passed around in manuscript among convivial friends. Even Williams & Wilkins seem to have been doubtful about it; for, although they had wanted it in a hurry, they sent it to a number of "readers" before issuing it a year after Bell had submitted it. Such delay was unusual in his relations with that publisher.

"Time after time, with a serene forgetfulness of earlier wrecks," Bell explained in the introduction to the book, "our hopeful species has donned a lifebelt as the ship was about to founder, confident that this or that lifesaver was less rotten than its predecessors. And time after time the cupidity or incompetence of the manufacturers of lifebelts has let down the deluded believers who trusted them. If the manufacturers of salvation have let us down, we, not they, are the more culpable . . . ; for we, not they, have been crass enough to believe that this time, surely, the new lifebelt will sustain us."

He then took up various "lifebelts." Some were designated as being of different eras—Religion I as well as Religion II and III and Science I, II, and III. Other lifebelts were Democracy, Industry, Invention, Economics, Psychology, and Education. He concluded with "The Lifeline"—Science III.

The book was dedicated "To Grace, whose fireside chats on Life are to blame for this book." Although Mrs. Hubble declined to have her last name included, she played the kind of active, personal role in the book that "Toby" played in *The Search for Truth.* At the end of Chapter X, however, Bell had to take leave of her before inspecting the last lifeline—Science III:

"She refuses to accompany me because she has already expressed her opinion that 'science is not going to get the public anywhere.' I am interested in trying to see whether science is going to get itself anywhere. Wherever it gets, there, I believe, the public will also get."

Not willing to let the subject lie—surely another fault in the book—Bell proceeded to list as postscripts a few (in fact, forty-five) "lifebelts" he had passed over, adding Grace Hubble's supposed comments:

4. Coeducation. Might have floated if there had been four sexes instead of two.
5. More free public libraries. Ask Carnegie.
29. Monarchy. Is one fool better than none?
31. Bureaucracy. Too much like a chest of drawers.
40. Do unto others . . . Overweighted with woulds and shoulds.

Reactions of reviewers were almost unanimously negative.

"Professor Bell regards himself as a toughminded realist who considers "good" and "evil" as subjective, yet he writes with constant moral heat and from a most casual acquaintance with facts."

"Dr. Bell, whose keen intelligence is unhappily subject to tantrums, rides his pet aversions with all the feckless rage of Don Quixote . . . ."

Only the *New Yorker,* the pages of which had seen such reviews as "Tonstant Weader fwowed up," found the presentation "fresh, witty, and even amusingly crotchety" although "the ideas are not novel."

The following summer Bell whipped off another book: *New Magic for Old,* subtitled *Science in a Dawning Civilization,* between June 23 and August 6. It was never published, being largely a rehash of much he had already written.

That same summer of 1938 the American Mathematical Society celebrated the fiftieth anniversary of its founding. In addition to the featured speech by Birkhoff on "Fifty Years of American Mathematics," there were to be talks by other mathematicians on developments in specific fields. There was no scheduled talk on number theory, but Birkhoff mentioned "the stimulating work of E. T. Bell" as well as that of "a very active and able group of younger men ... who are obtaining valuable results." Among these were a number who had received their inspiration from Bell.

Bell, who spoke on "Fifty Years of Algebra," announced that he was going to limit himself "to the broader aspects of the rapid growth from the age of relative algebraic innocence, when everything was special and detailed, to our present highly sophisticated abstraction ....

"[N]either the special nor the general has been solely responsible for the progress of algebra, in either division, at any time in the past, and there is at present no evidence that one can be cultivated independently of the other if algebra is to continue to grow. The interest here in this interplay between the particular and the general is that in the past third of a century the trend has been increasingly away from the particular ...."

The third of a century of which Bell spoke had spanned his mathematical life, and in these years he was trying to cope in his work with the changes the trend to abstraction had engendered in algebra.

"I think Bell felt he *ought* to be interested in this abstract side of algebra and to follow it," I am told by Dilworth, who by 1938 was a graduate student of Ward's and "interacting" as well with Bell. "For example: rather than looking at the integers, which form a semi-group under multiplication, he got interested in to what extent some of the number theoretic problems and number theoretic ideas could produce very general results in, say, commutative semi-groups. Well, a general semi-group is a long way from the integers.

"The main difficulty was that there were certain kinds of techniques that Bell found very profitable, techniques he liked to apply and so forth. He felt very much at home with these, so when he sat down to work in any area of number theory—diophantine equations or paraphrases or whatever—he already had at his disposal a lot of tools. In abstract algebra he didn't have any of those tools."

Among Bell's mathematical tools was one that he had used for a long time in his work on special sequences and formal power series. This was the umbral calculus, which he had first met as the symbolic method in Lucas's *Théorie des nombres*. During his career he was to write three general papers on the umbral calculus, the first having been "An algebra with a singular zero" back in 1923. The second was a historical and mathematical article in 1938 in which he pointed out that the Rev. John Blissard (1803–1875) had introduced what he called *the representative notation,* but which was actually the same thing as the symbolic method, "fully developed" and "in a mathematical journal with a wide circulation," fifteen years before Lucas. The latter, however, had always been given credit for its invention. Fairness and justice were important to Bell—Arnold emphasizes this aspect of his character, and the purpose of his second paper on the subject was to set the record straight. His third paper, in which he attempted to establish a sound postulational foundation for the subject, was not to appear until 1940.

The umbral calculus, to which Bell was so extremely attracted, is in a way a kind of *paraphrase.* The leading and most enthusiastic exponent of the method today is Gian-Carlo Rota of the Massachusetts Institute of Technology, who explains it for my edification as follows:

"Originally, you see, it was noticed that when you have an A sub n sequence, computations can be made easier if you treat the index as if it were a power. But then it becomes an art to know when not to treat it as a power and when to treat it as a power. [Laughing.] And that's the source of the umbral calculus."

It is a tool of which many mathematicians are highly suspicious. "A very shadowy subject," some murmur, making a play on the name. Even when they obtain results with the umbral calculus, so I am told by Rota, they will write up their results as if they have been obtained by standard methods.

I ask Rota if there are any results that can be obtained with the umbral calculus that can't be obtained by standard methods.

"No." Flatly.

"Then what do you get from the umbral calculus?"

"You get insights."

I am interested in how Rota, who never met Bell and knows nothing of his personal life or of his nonmathematical writing, happened to become interested in his work several years after his death.

"In 1963 I was then changing fields, changing from analysis to combinatorics," he explains, "and once a week John Riordan and I would meet for lunch. He was still working at the old Bell Labs, and I would pick him up at that old nineteenth century building. He himself looked like a character out of the nineteenth century, always wearing a hat, which was very quaint at the time. And he would take me to these extremely unusual restaurants. But expensive. And we would talk a lot about the material in his book, *An Introduction to Combinatorial Analysis,* where he quotes Eric Temple Bell extensively. [Riordan's book is in fact dedicated to Bell.] It was a very influential book. Very influential on me, but also very influential on many people."

Today Rota and a group of colleagues are attempting to establish a rigorous foundation for the umbral calculus, something that Bell was trying to do in his 1940 paper "Postulational bases of the umbral calculus." He sends me three pages of the new work—"The Classical Umbral Calculus," written with Brian Taylor.

"I believe the present version is at last the rigorous one," he writes, "and, believe me, it has taken us more time than it should. The applications of the calculus . . . are much the same as Blissard, Bell and Riordan envisaged; however, the present version makes their calculations completely rigorous, or so we believe. What makes the umbral calculus so odd from the point of view of contemporary algebra is the fact that certain identities are valid only if the letters occurring in them are distinct. This is one reason why these letters are called umbrae rather than variables. Such an assumption is, to the best of my knowledge, without precedent in the development of algebra since Hilbert. It is,

AT&T, Bell Laboratories Record

JOHN RIORDAN
a dedication to Bell

however, not without precedent altogether: a similar situation occurs also in electrical circuit theory . . . . Once one admits the postulate that substitutions can be made only if the variables are distinct, then the umbral calculus loses its weirdness, though not its fascination."

Weirdness. Fascination. Rota, having just finished teaching a course on the umbral calculus, confesses as well that "there is still an air of *witchcraft* about it." His words intrigue me. I have always wondered if there is anything in Bell's mathematics as bizarre as some of the things in his science fiction. Perhaps from a mathematician who speaks of "witchcraft" I can get some clue as to how the same man could write *The Scarlet Night* and *The Crystal Horde* as well as some two hundred mathematical papers. Rota has not known that Bell wrote science fiction until I tell him, but he answers promptly as to Bell's mathematics.

"It's *bizarre.* There are other mathematicians that are bizarre. Sylvester is equally bizarre, but he is a much stronger mathematician. There are nowadays mathematicians who are making their careers rewriting some papers of Sylvester's. You cannot say that of Bell. But Bell is in the line of the eccentric mathematicians. Like Sylvester. The mathematicians who break with the accepted tradition, who have a liking for the other side of mathematics."

(It should be noted that when I ask Dilworth the same question, he dismisses it out of hand: "Bizarre mathematics just isn't good mathematics." Durst has a similar reaction. They are not of the eccentric tradition of mathematicians.)

By the time that Bell was trying to put the umbral calculus on a postulational basis, his mathematical productivity had begun to taper off. In the five-year period 1927–1931 he had averaged more than ten papers a year. In the subsequent seven-year period 1932–1938 the average dropped to a little over seven. There were still to be a couple of dozen papers in the fields of his earlier mathematical work, but his interest in these fields had diminished. More and more he concentrated on multiplicative diophantine analysis, where he was to produce significant papers up through 1947.

He appears to have begun to have a sense that he was winding up his research career. In 1938, after he delivered his talk on "Fifty Years of Algebra," he arranged to have reprints of all his mathematical papers to date collected and bound. The three volumes, now in the Institute Archives at Caltech, are each inscribed "February 7, 1939"— his fifty-sixth birthday. Examining the volumes, I am surprised to find there a paper from the *Annals* entitled "Chains of congruences for the numerators and denominators of the Bernoulli numbers" that is not listed by Durst in his bibliography of Bell's work. Then I see that the paper is not by E. T. Bell but by Jessie L. Bell.*

Taine is more surprised than I, since he has never known that his mother took a bachelor's degree in mathematics at the University of Washington before he was born. I find one other case of Toby's mathematical participation in her husband's work. In "Exponential numbers" she is credited with having calculated the accompanying table.

The same spring in which Bell inscribed the three bound volumes of mathematical papers, John Taine resurfaced in science fiction for the first time since 1935.

"Scientific fiction is coming back—editors are asking for it," he reported happily to Hughes. In the same letter he reported that he had been approached by McGraw-Hill, a publisher with whom he had not dealt before, with a proposal that he write a history of mathematics to be used as a textbook. "Like a chump, I agreed." He signed the contract on January 21, 1939, and promised that the manuscript would be submitted to the publisher *within a year.* He himself expected—he told Hughes—that he would be "up to his neck" only until September.

There was a short story by John Taine, titled "The Ultimate Catalyst" and billed as "A Novelet of Super Chemistry," in the tenth anniversary issue of *Wonder Stories.* According to an accompanying note, Bell liked the story "because its weird pure s-f chemistry shocked my friends among professional chemists to the soles of their boots. Being an academic man myself, nothing gives me greater pleasure than to take some of the excelsior out of over-stuffed shirts."

He also had a guest editorial ("Why Science Fiction?") in *Startling Stories.* Entire nations were currently being led by men who feared and hated science, he pointed out. It was necessary that the public recognize for its own good that science was worth getting acquainted with: "Writers of science fiction need not preach, or even teach deliberately. If they tell a good story well, they will have done their bit."

Tomorrow, the novel Bell had written after spending a night with Hubble on Mt. Wilson in 1928, appeared in *Marvel Science Stories* in 1939. This novel, like *Twelve Eighty-Seven,* has never appeared as a book and is dismissed in the *Science Fiction Encyclopedia* as inferior, "a thriller that turns into a DISASTER story."

---

* The paper gives a number of congruences relating numerators and denominators that are useful for checking the accuracy of computed values.

"That's really inadequate and misleading," Douglas Robillard, a retired professor of English at the University of New Haven, writes to me. In his opinion, the novel is one of John Taine's more interesting ones. "Of greatest interest is the basic theme, mirrored in the title. *Tomorrow* is about a quite distant future. This emphasis is unusual for Taine, and he is able to make a lot of it. The story is . . . something to hang the reader's interest on while Taine teaches him that nationalism is the leftover disease of the 20th century to be cured only by 'one world' cooperation. We get the thoughts of the author on the last page when the scientists, who have been able to discover the location in the galaxy of the richest source of the ray they want to study and use for the (genetic) betterment of humanity, are preparing to begin their experiment with six hundred volunteer married couples—tomorrow."

In 1939, when *Tomorrow* appeared, Bell was not giving John Taine much time or thought. Europe was moving inexorably toward another war. Toby was again not well. In a dozen or so letters to Taine, who was driving across the country for a summer at Woods Hole, she constantly tried to reassure him. "My blood pressure and heart just needed a rest." She was feeling fine and would "probably only have to spend two or three days in bed." Romps was being "an angel," washing dishes and making his own bed. "I am getting steadily better." The reports on her health were accompanied by references by herself and Bell to *The Development of Mathematics* (never named, however), the book he had agreed to do for McGraw-Hill. Toby wrote that Romps was working on it "and loathing it." Bell contributed that he was a little more than "up to my ass with the g.d. book." Romps found working on the book "irksome," Toby wrote to Taine, and, later, Romps was "still digging away at the book." Also running through Toby's and Bell's messages to Taine during the summer of 1939 were complaints about "the Greek" and problems in the mathematical faculty.

By this time Bell had got Millikan to offer Morgan Ward a position and a salary that would keep him at Caltech. Ward was a man of many talents and interests, almost universally liked and admired. Apostol, as the longtime chairman of the committee that now awards the Morgan Ward Prize to the winner of a mathematical competition limited to freshmen and sophomores, describes him to me as "an eminent number theorist, an accomplished pianist, a serious student of modern poetry, an authority on succulents and cacti, and a champion GO player." In his way, according to Dilworth, who was his first Ph.D., Ward was as complex a personality as Bell.

The two men, once teacher and student, worked easily together. When Dilworth decided that he was more interested in lattice theory, about which Ward was talking with enthusiasm, than in Bell's idea of seeing what could be done with number theory in the environment of semi-groups, he found that "E. T. didn't mind a bit that I switched over." But the addition of Ward to the faculty divided the mathematicians—with Bell and Ward on one side and Michal, weakly but faithfully supported by Bateman, on the other.

"Bateman wasn't really happy about the situation," Dilworth tells me. "He was a very gentle soul. He just wanted to get along with everyone, so it was a very embarrassing and awkward thing for him. One of the reasons he lined up with Michal was Mrs. Bateman. She was a delightful person but a very stuffy Englishwoman. I think E. T. was genuinely fond of Ethel Bateman, but he just couldn't resist poking fun when he thought she got a little pompous. Of course Harry Bateman didn't particularly appreciate that. On the other hand, he was not the kind to come back with force."

In 1939, with Michal's promotion to a full professorship, there was a definite need for a new man "of large promise" (so Millikan put it) to improve the mathematics teaching of freshmen and sophomores and to meet the applied needs of the Institute.

"The Chief's way of operating in those years," Dilworth explains, "was to call in each member of the mathematics faculty and see who they had to recommend. Well, what happened was that Millikan would call in A. D. Michal and he would have certain suggestions— in fact, A. D. was very anxious to get some of his students back, students who had finished their work and gone elsewhere. Then Millikan would call in E. T. Bell, and E. T. would absolutely turn down the suggestions of Michal and have some suggestions of his own. Michal in turn would veto those. So for a long time they made no progress at all in adding anyone to the department."

A visitor at Caltech during the academic year 1939–1940 was the Swiss-born mathematician H. F. Bohnenblust, who was on a sabbatical leave from Princeton. Bell had great respect for him and for his wife, Eleanor, and he asked them to read *The Development of Mathematics* in manuscript. They recall Toby resting on the couch while they talked with Bell about his book. It was clear that he was worried about her.

"He was very dependent on her, but not in the helpless way many mathematicians are dependent on their wives," Eleanor Bohnenblust

LUDDYE MICHAL at the time Bell first met her in 1924 at Chicago

says. "She gave him vitality, self-confidence, adventure, challenged him to do things, took him places, put starch in him."

I have no accurate knowledge of the chronology of Toby's final illness, but in a copy of Walter De la Mare's *Behold the Dreamer,* which she gave Bell on his birthday in 1940, there are several penciled notations. The first is "May 4, 1940," when she may have entered the hospital. She seems to have been home again for a while but then to have gone back into the hospital in September. Bell wrote down September 11, 1940. Nine days later he wrote 7:21 p.m./September 20. She died on September 22.

The Development of Mathematics appeared the following week.

Bell would see no one but Luddye Michal, who came and did those things that had to be done, sorting Toby's belongings, packing up her clothes, putting away her pictures. He wanted no observance of her death. No funeral or memorial service. No flowers. There was nothing in the local paper about her passing except the official listing of name and date of death. He gave instructions that she was to be cremated, and he took possession of her ashes. At Easter he planned to go up to Yreka and scatter them by the big rock outside town where they had so often sat and talked during their courtship, and where he had asked her to marry him.

FIVE

the low road

September Song

The effect of his wife's death on Bell's productivity is evident in his bibliography. Only two mathematical papers appeared in 1941, a mere dozen pages. There were no book reviews and no general articles. No popular books on mathematics. No science fiction. It was six months before he could speak of her even to so close a friend as Grace Hubble.

He escaped his empty house by dining at the Athenaeum, frequently during the spring semester of 1941 with A. W. Tucker and his wife at that time, now Alice Beckenbach. Tucker was on leave from Princeton to work on a history of combinatorial topology. In the course of their conversations, he discovered that, although Bell disclaimed any real knowledge of that subject, he knew much more than he admitted. Talk regularly continued until midnight when Tucker, realizing that Bell hated to go back to his empty house, would walk him home.

The Development of Mathematics still strikes Tucker—among books on the history of mathematics—"as the most interesting as far as I am concerned." Unlike *Men of Mathematics*, which he finds "almost fiction," *The Development of Mathematics* was intended for an essentially professional audience. Although Bell had hoped to find among his readers some undergraduates wanting to extend their education beyond the calculus, he had directed his book primarily to graduate students and young instructors. Many of these, he thought, would like to have a broad treatment of their subject, emphasizing concepts and methods that had survived and including "technical hints" as to why certain things continued to interest creative mathematicians while others did not. Such a treatment might enable them to decide on the particular field to which they wanted to devote their efforts and (here one detects a dig at Michal, and perhaps even at Blichfeldt) to resist the blandishments of teachers trying to direct them to their own specialties. His history would differ from others "hallowed by tradition" just because he was a mathematician. A historian "tended to emphasize the smoothness of the historical development" while a mathematician was more likely to see the "discontinuities," realizing that "what looks like an anticipation after an advance has been made may not even have been aimed in the right direction."

Contemporary historians of science were on the whole approving of Bell's foray into their field. Although I. Bernard Cohen found many statements about mathematics before 1637 (the date of Fermat's Last Theorem) "passed out in the grand manner which may irritate historians," he added that for the rest "Mr. Bell has given us the best concise account of the development of mathematics in any language." The emigré historian of ancient mathematics Otto Neugebauer, reviewing a German paper on histories of mathematics, chided its authors for omitting "such an outstanding work as E. T. Bell's *Development of Mathematics*, [which] represents perhaps the most successful attempt at extending the historical narrative down to modern times."

Even in such a serious, scholarly work, however, Bell was still Bell. Index entries as well as text reflect his personality and prejudices. Under "God," one finds "R. Bacon's, G. Boole's, S. Clark's" and so on to "Plato's." "Miller, G. A." is indexed, but on the cited page there is no mention of Miller (from whom Bell took his first course in the history of mathematics at Stanford), only a diatribe against Miller's mathematical specialty, "this troglodytic sort of activity." Miller, who was fond of writing articles with such titles as "Twelve Mathematical Errors in Webster's Dictionary," was long rumored to have sought out every copy of *Development* at the University of Illinois and to have noted in each errors Bell had made. A later Illinois mathematician, Bruce Reznick, located one of these, which listed twenty-one errors and exaggerations, including some labeled simply "bad" or "silly."

(Reznick, B.S. Caltech '73, was the recipient of both the Morgan Ward Prize and the E. T. Bell Undergraduate Mathematics Research Prize, awards that were established in 1963 to keep alive the names of Bell and Ward. The list of recipients of both prizes is impressive, according to Apostol, who has long chaired the Awards Committee.)

During the Easter vacation following Toby's death, Bell went to visit Taine in San Francisco, taking with him her ashes to scatter in Yreka. Taine had just met a young woman named Vicki Stocker who

RECIPIENTS OF THE E. T. BELL UNDERGRADUATE MATHEMATICS RESEARCH PRIZE			
1963	Edward Bender	1980	Eugene Loh
	John Lindsey II		John Stembridge
1964	William Zame		Robert Weaver
1965	Michael Aschbacher	1981	Daniel Gordon
	Richard P. Stanley		Peter Shor
1966	(no award)		Thiennu Vu, first woman
1967	James Maiorana	1982	Forrest Quinn
	Allen J. Schwenk	1983	Mark Purtill
1968	Michael Fredman		Vipul Periwal
1969	Robert E. Tarjan	1984	Bradley Brock
1970	(no award)		Alan Murray
1971	(no award)	1985	Charles Nainan
1972	Daniel J. Rudolph	1986	Arthur Duval
1973	Bruce Reznick		Everett Howe
1974	David S. Dummit	1987	Jonathan Shapiro
1975	James B. Shearer	1988	Laura Anderson
	Eric D. Williams		Eric Babson
1976	John Gustafson	1989	James Coykendall IV
	Albert Wells Jr.	1990	(no award)
	Hugh Woodin	1991	Allen Knutson
1977	Thomas G. Kennedy	1992	Robert Southworth
1978	(no award)		Michael Maxwell
1979	(no award)	1993	(no award)

taught at a private school in Pasadena and was in San Francisco for her spring break, and he invited her to accompany him and his father on their trip to Northern California. Now the wife of Leverett Davis Jr., Emeritus Professor of Physics at Caltech, she recalls how a little outside the town of Yreka she and Taine sat in the car while Bell, carrying something—a small bag or box, she cannot remember—walked up a hill to an outcropping of rock with a tree beside it. Afterwards, on the way back to town, he pointed out landmarks to Taine and later drew a map: "Directions for locating spot where Toby's ashes are, and where mine are to be scattered."

Upon Bell's return to Pasadena, he kept up his acquaintance with Taine's friend, and she introduced him to a Wellesley classmate, Nina Jo Reeves, who was working for a Hollywood studio. Both Vicki and Nina Jo had been English majors, and they marveled at a mathematician who knew about literature. Nina Jo had a car, and they invited Bell to join them on drives "sightseeing Southern California."

"People must have found it a kind of funny thing, I suppose," Mrs. Davis says today, "but that never entered our heads. He was such an *interesting* man."

At the beginning of the summer of 1941, Bell turned again, apparently for the first time since Toby's death, to "recreational writing"—not science fiction, but a revised version of *The Search for Truth*, which his friend Zeitlin wanted to publish. The year of the publication of the earlier book (1931) had seen the appearance of the landmark paper by Kurt Gödel that had dashed David Hilbert's hopes of establishing the consistency of mathematics. Gödel had done this by showing that it is not possible in certain logical systems (including *arithmetic*) to prove, by the rules of the system, certain theorems belonging to the system which can be seen otherwise to be true.

"This is no mere existence theorem of the non-constructive type . . . ," Bell wrote in *The Development of Mathematics*. "Gödel constructed a true theorem such that a formal proof of it leads to a contradiction. Undecidable statements exist: within the system certain assertions can be neither proved nor disproved."

Undecidable statements exist. This was an idea as enthralling to Bell as the idea of multivalued truth. When he had been writing the earlier book, Gödel's revolutionary result had been too recent to have been passed upon by mathematicians. Now, Bell felt, it was time to go back to the subject of *truth*. In a handwritten introduction to the new book, which he intended to call *Nothing But the Truth*, he again brought in the character of Toby, "who is no longer here to defend herself."

Bell completed *Nothing But the Truth* in the summer of 1941, but it was never published because by December of that year the United States was at war.

As an all-male educational institution with a highly specialized technical faculty, Caltech was even more affected than most colleges and universities by the all-out military effort of the nation. But except for the change in the type of student and the necessity for teaching on occasion during the summer, Bell was never as involved as most other faculty members in "war work."

By the late spring of 1942, he was finally getting his personal life back together. Grace Hubble, picking him up for a dinner party at her house, commented on his "looking nice, very well groomed, in a dark blue suit." A few weeks later, invited to his house for tea, she noted that he had cleaned the kitchen and got out the best teapot in her honor.

"We talk of the lack of continuity of a personality, that is, can you think of yourself at 7, 17, 27, etc., or aren't you a lot of other people?" she reported. "Of loving anybody later in life, & Romps says it's like the title of a book by Thomas Wolfe, 'You Can't Go Home Again.'"

That fall Bell accepted an invitation to spend two months at the University of Kansas City (now the University of Missouri at Kansas City). There he met a Mrs. Leoda Marie Stout, a handsome longtime divorcée, thirty-seven years old, who was giving up her successful wholesale cosmetics business because of ill health. Not long after their meeting, Mrs. Stout and her seventeen-year-old daughter, Rosemary, moved to California.

By this time, although Taine had been admitted to candidacy in biology at Stanford, he had come to realize what he now thinks he had always known. He did not like or want the academic life and would not be successful in it. To his father's disappointment, forcibly and often cruelly expressed—on occasion in the presence of others—he switched from biology to medicine.

"Romps was not very happy about Taine going into biology in the first place," Luddye Michal told Albers during his interview with her in 1981. "Of course later, when Taine decided to go to medical school, that was the absolute bottom. You see, first he could be a mathematician and failing that, well—he could be a physicist and, failing that, maybe a biologist. And he should be a writer too. Being a doctor—well, he couldn't sink much lower than that."

In medical school at Stanford, Taine met a young woman named Janet Snelling about whom he was soon to become serious. It was she who in the fall of 1943 began to save the letters from his father that I am quoting in this and subsequent chapters. Rather surprisingly, in view of the father-son relationship, the letters are quite ordinary communications from a parent to a grown child. News about people Taine knows and friends of his that Bell happens to see. Advice about "clipping coupons," turning in bonds that are due. Warnings to eat well, get plenty of sleep, keep "regular." The only indication of Bell the writer is in his vivid descriptions of the activities of various cats in the neighborhood. There is virtually nothing about mathematics or about what is going on in the mathematics faculty.

Michal had pushed hard for a former student of his, and he was bitter when "the Chief" decided to approve the Bell–Ward candidacy of Dilworth. After another year at Yale, brushing up on statistics, Dilworth joined the faculty in 1943.

R. P. DILWORTH, student, colleague, and friend

"After I came back, in the meetings to determine departmental policy and things of this sort—it was then always three to two," Dilworth tells me. "There were lots of departmental things that had to be settled, since they had not been settled before because the vote was always even. I changed the picture. We had some very lively meetings, let me say, at that time. And it was a very tough time for A. D. He was just taking one beating after another. E. T., on the other hand—after putting up for such a long time with Michal and his ways—just couldn't resist using some of these meetings to, if you wish, take advantage of A. D.'s position. There were occasions where he would overstate things a bit just because they would be a little more effective that way than otherwise, but not in the sense of lying—let's say *overemphasis* in some respects. So he essentially enjoyed the situation at this time. On the other hand, with Morgan Ward it was just the practical aspect—now things were going ahead—we were getting things done—and so on. He wasn't taking pleasure in the situation. As for Bateman, he would sit and hardly say anything. But he would vote with Michal."

Dilworth had returned to the Institute with a bride, and Bell invited her to tea at his house.

"E. T. had eyes for the women. There's no doubt about it. And they would just sparkle!" Miriam Dilworth says. "But he wasn't fresh with you. He could be crude, and then look to see if you were getting the point, but he was never forward—to use an old-fashioned term. I went to his house for tea, and I was the only one there. It was perfectly all right. There was nothing like what we now call 'sexual harassment'— and I was at an age when that was a distinct possibility."

It was not until the following year (1944) that Leoda Marie Stout began to appear in Bell's letters to Taine. In July he asked "for a friend" about a good cardiologist in Los Angeles.

"The reason I want to know is that she has asked me to look out for her kid, in case she passes out unexpectedly, until the family lawyer in Kansas City can take hold. I'm to see to the cremation, &c. She is convinced she hasn't much longer, and I have a hunch she is right . . . . I will have no responsibility . . . except to see that her wishes are carried out, & to take care of the girl (age just 20), who will be an orphan with a considerable fortune, till things get straightened out . . . . For some reason she sized me up as being reliable for the job."

There was nothing more about Bell's sick friend until the beginning of September when Taine wrote that he would be coming to Pasadena later in the month.

"Is Janet likely to come too?" Bell wrote back worriedly.

Sometime earlier, his house had been enlarged. Now, he explained to Taine, he had moved out to the new "sleeping porch" and had rented the two bedrooms to Mrs. Stout and her daughter.

"[They] were desperate for a place, and could get nothing except this. It would have been inhuman to refuse, and I expected Mrs. Stout to have been able to move out by the 17th of Sept.* So you will have to make it in your sleeping bag."

For Taine it was as if he had been thrown out of his own home, but for Bell the new circumstances were welcome. Suddenly he found himself with a family again, a family with the delightful addition of a daughter.

From this point on, Bell's letters to Taine regularly reported his developing relationship with the Stouts. As was his custom, he gave them nicknames. Mrs. Stout was "Cagey" or "Cagey Cat," sometimes simply "K. G.," because, according to her daughter, "she played her cards very close to her chest." Rosemary herself was "Pack Rat" because of her habit of saving clippings for the art classes she was taking.

Bell's letters were filled with reports of emergencies, trips to the hospital, relapses. His guess was that Mrs. Stout did not have "much longer to go." When she was confined to bed, he took Rosemary to dinner at the Athenaeum.

In an interview with Albers in 1981 Rosemary Stout Swift recalled Bell as being as much of a father as she had ever had.

"One of the things I always felt about him was that I could trust him. Even when I was that young. Somehow I felt that he would not ever mislead me in any way. I don't think many people felt that way about him. But I could always get a real honest answer from his point of view. He was such a definite person. There was nothing wishy-washy about him. You really had to make up your mind right away whether you could stand him. I took to him right from the beginning. Even at seventeen, when I first met him, I thought he was great."

Bell's letters to Taine during this period also enable me to date an event in Bell's life that is legendary at Caltech.

"Yesterday evening, there was an open forum sponsored by the Y.M.C.A. on the Existence of God," he wrote on September 1, 1944. "Pa

---

* Housing was very scarce in many parts of the country during the Second World War, and especially so in Southern California, where there was a large defense industry.

LEODA MARIE STOUT, "excellent company and a witty conversationalist"

Millikan roped me in to debate with Dr. Blake of the First Presbyterian Church."

Many people recount to me the story of the debate between Bell and Dr. Eugene Carson Blake, a distinguished liberal clergyman who later became even better known for his support of the Civil Rights Movement. They describe how, rather than trying to prove his own case, Bell constantly pressed Blake to state the premises from which he derived the existence of God. They also describe how some students rigged up a contraption with a lighted bulb attached and lowered it from a trapdoor above the lectern while one of them demanded in a sepulchral voice, "If I am not God, then who am I, Dr. Bell?" But not one of those recalling the event had actually been present at the debate. I finally write to Kurt Mislow, the Director of the Frick Chemical Laboratory at Princeton, at the suggestion of Laura Marcus.

"I vividly remember the moment you refer to in your letter," he replies. "As the voice spoke (I don't remember the exact words, but what you quoted sounds about right) a lighted electric light bulb descended slowly from the ceiling above Bell's head, symbolizing, I suppose, a Ghostly Higher Being. Everybody in the audience (including myself) was convulsed with laughter!"

Durst, whom I have neglected to ask about the debate, notes when reading my manuscript: "I was there, too. Bell's answer to the question, 'Then who am I?' was characteristically from scripture:—'In your own words, "I am what I am."'" (1 Cor. 15:10) Some present may have thought he was quoting Popeye the Sailor!"

Bell himself did not refer to the students' prank in his letter to Taine, but only to the fun he had had responding to hecklers and to the size of the crowd. He had expected only a dozen or so, but more than four hundred people had turned up.

At the beginning of 1945 Taine and Janet, both by then with their medical degrees, informed Bell that they were going to be married that spring at Janet's home in Penryn, a little town in Northern California. He immediately began to plan his trip to the wedding. Although it was still wartime and travel was difficult, he was determined to spend some time in San Jose. People there had asked to see him, he wrote in one letter—and in another, "I shall probably stay in S.F. for two or three days anyway, & will then go down to San Jose, & may make my h.q. there."

It is hard to imagine how Bell, a rather well-known man whom people probably took pride in knowing, could have remained in contact with those who knew him when he was growing up in San Jose but could have kept secret from everyone else the fact that he had ever lived there. Yet the only hint of any knowledge of his connection with San Jose comes to me from Robertson's daughter, Marietta Fay. In her teens, she had a close and friendly relationship with Bell—one that in recollection surprises even her, for she still feels a bitterness about what she considers his betrayal of her father. After Toby's death, since he didn't drive, she would take him "to clip his coupons." Afterwards he would take her to lunch—for fish, which he loved. Other times she and George McManus, a CIA man who was living at the Robertsons' home, would take him "bar hopping." Since Marietta was only sixteen,

LINCOLN K. DURST as an instructor at Rice Institute in 1952

McManus enjoyed frightening Bell about the possibility of "a raid" and instructing him in detail as to how he should behave in that case. It is in the middle of our lunch at the Athenaeum that Mrs. Fay, recalling these events, suddenly interjects into our conversation, "What about San Jose?"

I am startled.

"How did you happen to say that?"

"I don't know," she says. "I *really* don't know."

I probe further but end up with no indication that she has any knowledge of Bell's youth in America.

With Taine and Janet looking forward to their wedding, Bell began to think about marrying Mrs. Stout. His young people discouraged him. He would be marrying a woman who was seriously ill—did he really want to take on such a burden? Remembering the trauma of Toby's long dying, he gave up the idea. Apparently the relationship remained platonic. Other faculty members were also interested in the attractive divorcée and Bell reported her as receiving at least two proposals.

Taine and Janet were married on March 29, 1945. Since Taine did not want to ask a classmate to make the difficult wartime trip up to

THE MARRIAGE OF TAINE AND JANET with Bell as best man

Penryn, he asked his father to serve as his best man. Years later, he and Janet were surprised to find among his memorabilia the withered carnation he had worn that day in his buttonhole.

In Bell's subsequent letters to the newly married couple, V-E Day, the dropping of the atom bombs, V-J Day, all passed with brief references or none at all. The Packrat celebrated her twenty-first birthday. Cagey's health continued to improve—her doctor brought in other doctors to show her off as "his prize specimen." She planned to find another place to live as soon as Rosemary finished school.

Bell was writing again. The first installment of a series called "Sixes and Sevens" had begun to appear in *Scripta Mathematica* in 1943 and was to continue until 1947. It was published in 1946 as *The Magic of Numbers*. There was a second edition of *Development of Mathematics*, with fifty additional pages "updated to 1945," although (as Saunders Mac Lane pointed out) the "updating" on occasion consisted only in changing 1940 to 1945. Mac Lane, nevertheless, found in "this magnificent, inclusive, and provocative survey of the origin and adventures of mathematical ideas," the great virtue "that it does not merely record facts, but it arranges ideas and passes judgment as to their importance." He conceded that given the "tremendous scope" of the work there were bound to be errors "both of fact and of judgment," but he found the book a joy to read just because it provided "so many chances profitably to disagree" with its author.

"Is Plato as vicious as Bell's everywhere dense cracks would indicate? Does Bell overemphasize the importance of lattice theory and miss some of the significant developments in modern topology? Has this hard-headed author been duped by the advocates of Brouwerian logic and many-valued logic? Is Fréchet's work as significant as Bell claims? Might some mathematical war workers disagree with Bell's dismissal of spherical trigonometry as useless?"

The next year (1946) *The Time Stream* appeared in hard cover. Because the novel had made its only previous appearance in the pulps, it had never been reviewed. Thus the earliest contemporary reaction, except for the letters to the editor of *Wonder Stories* in 1931, came twenty-five years after its writing.

"*The Time Stream* is by all odds the strangest of all the John Taine novels," wrote P. Schuyler Miller in *Astounding Science Fiction*. "So far as I can recall, it was the first story to develop the now familiar concept of Time as a flowing stream into which one may plunge, to swim forward into the future, or back into the past.

"Rumor has it that it was [John Taine's] first novel, which may be true, for it lacks the craftsmanship and characterization of later books .... Even so, the Taine touch and the soaring Taine imagination are there .... There is a double impact [today] in some of the implied but never stated relationships between our own Earth, the dawn-world Eos, and the desert world of the far past."

Miller welcomed the news that the Buffalo Press, which had issued *The Time Stream,* intended to publish in hardcover all the John Taine novels that had appeared only in science fiction magazines: "There are few greater services they can perform for a generation of magazine readers who know not [John] Taine."

Because Bell had devoted a summer during the war to teaching V-12 students, he had earned a leave of absence. The result was that from June 14, 1946, to January 4, 1947, he would be completely free of academic duties. In his entire teaching career (thirty-four years up to that time) he had never had a sabbatical. Now he could look forward to six-and-a-half uninterrupted months of "intensive writing."

Before Bell's leave began, tragedy struck the little research group in mathematics. On January 21, 1946, Harry Bateman died, stricken by a heart attack on a train as he was on his way across the country to receive an award from the Aeronautical Society. Bell wrote an obituary for the *Quarterly of Applied Mathematics.* In it he recalled his longtime colleague as "an almost unique" mixture of erudition and creativity.

"It is most unusual for a mathematician to have the extraordinary range of exact knowledge that Bateman had, and not be crushed into sterility by the mere burden of an oppressive scholarship. But, as his numerous publications testify, Bateman retained his creative originality till his death."

He had also left much unpublished work.

"I got Sterling (historian, chairman of Faculty)* to promise to interview Mrs. B., and ask her to turn over all of his stuff temporarily to the Institute for safekeeping," Bell wrote to Taine. "Then we could get a couple of competent men, on Guggenheim or Rockefeller fellowships, to go through it all and decide what to do with it. Most of it is in shoe boxes, filed according to some system known only to God and Harry—anyhow to Harry."

---

* J. E. Wallace Sterling, later President of Stanford University.

Tom M. Apostol

ROSEMARIE STAMPFEL, a friend and typist for the "Bateman project"

(This suggestion of Bell's resulted in what became known as the "Bateman project," all three volumes of which were typed by one of his good friends on the secretarial staff, Rosemarie Stampfel.)

With Bell to be gone for six months, a replacement for Bateman had to be obtained promptly. In view of the contention in the mathematics faculty, Millikan appointed a search committee that did not contain a single mathematician. Its choice was Frederic Bohnenblust. Bell was delighted—he liked and admired both husband and wife.

I am not able to meet the Bohnenblusts in person, but I telephone them one evening. As to my questions about Bell, the two have somewhat different responses. Bohnenblust tells me that he liked Bell and always enjoyed talking him, but that he never felt he knew him: "He was not a man who had a need to explain himself to you. He was always behind the picture he presented." Eleanor Bohnenblust objects: "I think most women would say that they felt they knew him."

I mention the number of his women friends and suggest that it would have been nice if he had had a daughter.* She said that to him

---

* There is an unusual number of father/daughter relationships in the novels of John Taine, beginning with Vera in *Green Fire*.

once, Mrs. Bohnenblust tells me, and he agreed—"He said he would have named her Heather, which was his mother's name."

I point out that his mother's name was *not* Heather but Helen. A few days later I receive a note from Mrs. Bohnenblust, thanking me for a postcard reproduction of Pissarro's painting of Penybryn and adding, "As for E. T.'s faulty memory, I'm inclined to be indulgent and to see nothing more sinister than his often rampant fancy. *Heather* is more colorful than Helen. Spontaneous fiction! One of his talents."

Looking forward to his leave, Bell had already started to work on "the project" to which he intended to devote himself. Never mentioned by name in the letters to Taine, it was the revision of the epic poem, *The Scarlet Night,* that he had written in 1910. The spot he selected for his retreat was Redondo Beach, not too far south of Pasadena. Someone he called "the Angel Gabriel," but whom no one is now able to identify, transported him to a rented room near the beach. Although he returned from time to time to Pasadena, he worked quickly. A sample of the style of the poem, which is very long, will have to suffice here.

> Only this I know:
> Even as the living coals that glow
> Beneath the gray of ashes dead and cold,
> Still for a moment in their dying hold
> An image of the generative flame
> Extinct, so we who know not whence we came
> Nor what shall be our end, enclose a spark
> Of that supernal fire which cleft the dark
> And shattered it upon the primal sea,
> Which, like a serpent, girt eternity
> Before the stars were seeded through the vast.

"The job for which I came down here was done by Sept. 1," he reported happily to Taine and Janet. "It is now being read by competent judges. I will let it stew for a year."

Trips back to Pasadena, where the Stouts were still living in his house, became more frequent. He continued to report on Cagey. Apparently no longer such an invalid, she was writing a novel that he thought was "pretty good." She was going to get married again—"some prosperous guy [in Cincinnati] who has been after her for years," but before she left she wanted to redecorate his house "as a sort of payment for all I have done for them."

On November 1 he reported that the house was now "a knockout" —"a house in the modern manner"—"a showplace." His friend, Dorothy Diamond, who writes me from Australia, recalls a Chinese blue rug on the floor, red lacquer floor-to-ceiling bookshelves, a sectional sofa in canary yellow.

By then it was getting cold in Redondo Beach.

Bell decided that he would not return after he went up to Pasadena for the inauguration of Millikan's successor, Lee DuBridge. A few days before he was to leave, walking on the beach late one night as was his custom, he was knocked down, beaten, and robbed. He was brought back to Pasadena in serious condition. Mrs. Stout postponed her leaving for a week, and Rosemary sat up with him to see that he took his medicine.

The mugging on the beach was the second event, after Toby's death, which in the view of Bell's friends was to result in a considerable change in the man they knew.

E. T. BELL as he appeared in the year of his retirement*

Winding Down

It was now some twenty years since Bell had enthusiastically written to A. D. Michal about his plans for the development of mathematics "at Tech"—first to bring H. P. Robertson to Pasadena and then Michal himself. As his father's favorite poet could have told him, however, the best laid plans aft gang a bit a-gley. Michal had turned out not to be a good choice, as the long dissension in the mathematics faculty testified; and Robertson, when he did accept an appointment in Pasadena, accepted on the condition that he would not be a member of the mathematics faculty. But even in separate faculties, the relationship between Bell and his former student continued to be stormy, according to Jesse Greenstein.

"Oh, I'm sure it was," Dilworth laughs. "Everyone who associated with Bell had a stormy relationship with him. There's nothing unusual about that. But he and Robertson really had a deep affection for each other. I would say that many times during the week H. P. would be over visiting Bell.

"Morgan Ward, for instance, never went over like that, although Ward named one of his sons after E. T."

There is no question but that it was a pleasure for Bell to have Robertson and his family in Pasadena after 1947. A further source of pleasure at that time was a revival of interest in the science fiction of John Taine. In the fall of 1947, which was when Robertson arrived to take up his duties, Lloyd Eshbach's Fantasy Press published *The*

---

* Inscribed "With all good wishes to the ladies in the telephone room from that old S.O.B. E. T. Bell, June 1953."

H. P. ROBERTSON at the time he joined the Caltech faculty

Forbidden Garden. This was the psychologically convoluted story of the genetically "tainted" older brother exiled to India, which Bell had written during the Christmas holidays of 1928–1929. At the time the book was published, Bell's brother, James Redward Bell, was still in Ceylon, living in the Planters Association's nursing home at Welimada and suffering from what his son-in-law described as "old-age debility." He died two years later, a fact that Bell was to report casually at dinner with the Michals, adding that his brother had never married, "but he had probably left some bastards around."

The Forbidden Garden had been Toby's favorite of her husband's novels and, in exchange for her typescript, Bell made Eshbach a gift of his own handwritten manuscript, now among the Eshbach papers at Temple University. On a sheet of notepaper that he then placed in his copy of the published book, he summarized the circumstances of this rather odd exchange: the notations on the manuscript—the dates when he began and finished, the fact that he was in San Diego at the time—also, without comment, "On 4 Jan. 1929, Chapter XV there is the note 4.15 p.m. 33."

By the time *Garden* appeared in 1947, "John Taine" had written all the science fiction he was ever to write. There were, however, three novels to be published for the first time after that date: *Black Goldfish* (1948), *The Cosmic Geoids* (1949), and *G.O.G. 666* (1954). According to a "Faculty Portrait" of Bell in the Caltech alumni's *Engineering and Science,* the last named novel was written in 1940 but, since it took place in a dictatorship where the work force was being reinforced by mating human beings with gorillas, it was turned down by publishers during the war as unfriendly to the Soviet Union, by then an ally.*

Black Goldfish, which appeared as a novelette in *Fantasy Book,* was (apparently) the last science fiction story that Bell wrote, although not the last published. The black goldfish of the title is Cleo, a member of the domestic staff of a boorish "refugee" biochemist, who has given her that nickname because she reminds him of the sexy goldfish in Walt Disney's *Pinocchio.*

"In her own proper time and place she would have slit his midriff like an over-ripe melon," Bell wrote. "But being deprived of her birthright by accidents of time and place, she merely said 'Yes, sir.'"

---

* Everett Bleiler, however, recalls seeing in the early 1920s an item by "John Taine" in a copy of Hugo Gernsback's *Science and Invention* (a precursor of Gernsback's *Amazing Stories*), claiming that the "Bolsheviks" were mating human beings and chimpanzees.

It is likely that the inspiration for Cleo (who willingly joins in a plot to undo her employer) was Mrs. Frances Lemons, whom Mrs. Stout had hired in 1946 to cook and take care of Bell's house. Bell had given her a nickname too—she was "Black Bess"—but although he referred to her in his letters to Taine and Janet by the racial epithets that were as characteristic of him as his slang, he addressed her always as "Mrs. Lemons" and treated her with great courtesy. He described her to his friend Bullitt as "a very competent colored housekeeper who runs the place (and me too)," adding, "If her skin had been white instead of black she would have gone to the top."

By the time that *Black Goldfish* appeared, Bell was by no means so productive in mathematics as he had been in the past. He was nevertheless actively working on a number of things. He continued to indulge in what he described to Bullitt as "my vice, diophantine analysis," but most of his papers, although always containing some "real mathematics," tended to be expository. He revised a couple of the long poems he had written during his early years in Seattle, and wrote or revised some twenty-three mathematical articles for the new edition of the *Encyclopædia Britannica*. He also continued to write numerous book reviews.

On February 7, 1948, as he passed his sixty-fifth birthday, he began to discuss with McGraw-Hill a new and unusual book.

"Suppose that our atomic age is to end in total disaster," he was to write ultimately in the Prospectus. "What problems that our race has struggled for centuries to solve will still be open when the darkness comes down?" Most of the "great" problems of science were in his view either too ambiguous or too broad to treat as *The Last Problem*. His own nomination was a problem "that anyone with an elementary-school education can understand." His book would be "a biography" of the famous, unproved "Last Theorem" of Fermat, and a biography of Fermat as well.

In Bell's correspondence with Bullitt there is a two-page letter in which Bell states his personal opinion about whether Fermat had actually possessed the "the truly marvelous demonstration" of the theorem that he had claimed in the margin of his Diophantus. The letter was in response to one from Bullitt, asking if Bell agreed with G. H. Hardy, who like almost all contemporary mathematicians believed that Fermat had been mistaken when he said he had a proof.

Bell did not agree.

WILLIAM MARSHALL BULLITT in the early 1950s

"I think the balance of the doubt is in Fermat's favor," he wrote Bullitt on March 23, 1937. "First, it is often forgotten that Fermat was a mathematician of absolutely the first rank. As an originator of new and powerful methods he has no superior. Second, I believe he had an insight into the properties of the integers that has never been even distantly approached—even by Gauss."

Even by Gauss. These are shockingly strong words from a mathematician, and Bell repeated them in the course of his letter to Bullitt. But, as he went on to write, citing his old professor D. E. Smith's opinion in support of his own, "Fermat had, *in the theory of numbers,* the same sort of native superiority over the average *great* mathematicians, that a *calculating* marvel . . . has over the average child of ten."

Bell planned to begin his Fermat book in the summer following his sixty-fifth birthday, but in March he came down with a viral pneumonia. The illness caused him to think about his death and about the distribution of his estate, which was surprisingly large as a result of his own frugality and the good advice of a knowledgeable friend. During the war he had been concerned about his sister, Enid, who by then lived on the Isle of Wight and in the path of the almost nightly German raids on London. After the war (despite his claim that no one in his family knew his whereabouts) he regularly sent her food and clothing. Now he began to be concerned about her welfare after his death, and he arranged a $10,000 tax-free legacy for her in his will.

"It is another of these pesky business details . . . ," he wrote to Janet, who had become the member of the family with whom he communicated on financial matters. "I wrote to my sister, telling her [about the legacy] and asking her to accept it without enquiry of any kind, in case of my death. She agreed; and in turn told me of her will, with a somewhat similar provision.* Taine knows (as, I think, you do, too) about the underlined proviso."

Even today neither Taine nor Janet has any idea why Bell was so adamant that there be no contact between his family in America and his family in England. He had been so throughout his life, according to Taine, and he was to be so until his death.

By the summer of 1948 Bell was in reasonably good health again. He was working hard on *The Last Problem* and looking forward to at-

---

* Bell's legacy to his sister is confirmed in correspondence with his lawyer. There was, however, no such matching provision in Enid Bell's will when she died on February 2, 1959. In fact, the name of her only living brother did not appear in her will.

tending the dedication of the 200-inch telescope on Mt. Palomar, to which he had received one of the coveted invitations. (Grace Hubble quoted her husband as saying of the telescope, "We don't say it's perfect, because we don't use the word perfect. But it's almost a miracle.") When neither Taine nor Janet could go with him to Mt. Palomar, Bell asked Mrs. Stout. She had not gone through with her plan to marry her old admirer but had returned within a month to Southern California. Her health had improved remarkably, and she was holding down a job in Santa Barbara.

"I hardly expected her to come," Bell wrote Taine, "[but] she told the boss he could have her resignation unless he gave her leave."

He rented a car. The driver was "excellent," the weather was "perfect." Along the way he and she enjoyed a picnic lunch she had packed. He had been concerned about her heart because of the long walk up to the observatory and the many stairs but, except for a worrisome twenty minutes she spent in the women's room "powdering her nose," all went well. Among Bell's memorabilia there is a souvenir program inscribed "To Leoda M. Stout" with the comment that she is excellent company, a witty conversationalist, "and best of all knows when to keep her mouth shut."

Three months later, on September 2, 1948, she died at the age of forty-three.

He put her photograph away, but for the next few years he memorialized the date of her death in the volume of Robinson Jeffers's poems in which they had observed her Christmas birthday. As far as I know, the only deaths he ever memorialized were his father's, his wife's, and—for a few years—Mrs. Stout's.

He continued to work in the evenings on his book about Fermat and the Last Theorem, but progress was slow. Since Toby's death he had begun to drink more frequently. Some people felt that Mrs. Stout should have drawn a line in this regard, as Toby had, rather than joining him in a drink. ("But we should not emphasize the drinking," Bohnenblust says gently. "It is not a very pleasant subject.") Bell was capable of swearing off alcohol for several months at a time. Writing to Taine during one such period, he commented, "It is remarkable how [people] have stopped coming when they realize there is nothing to drink. Good riddance."

Rosemary Stout and her husband, Dean Swift, who had been one of Bell's doctoral students, suggested that he read *The Scarlet Night* over their new "wire recorder." He liked the idea; it was important with

MATHEMATICAL FACULTY 1948
Front row, left to right: S. Karlin, A. D. Michal, A. Erdélyi, Morgan Ward
Back row: R. P. Dilworth, E. T. Bell, H. F. Bohnenblust

poetry to hear how it sounded when it was read aloud. But when
he tried out the machine, he was horrified at the sound of his "G. D.
clipped British accent . . . [but] all said it really was my voice. My god!"

Rumor of *The Scarlet Night*, a new work by John Taine, circulated
quickly in the science fiction community. Eshbach wrote to Bell that he
would rather publish the "new" piece than *To Be Kept*, an earlier novel
that I have not been able to date. Although Eshbach subsequently
reprinted two John Taine novels as well as the previously unpublished
G.O.G. 666, neither he nor anyone else ever published *To Be Kept*.

In 1949 William L. Crawford's Fantasy Publishing Company issued *The Cosmic Geoids*.* Since it was relatively short, the volume was padded with *Black Goldfish* and titled *The Cosmic Geoids and One Other.* Bell had copyrighted *The Cosmic Geoids* in November 1942, and it is clear from internal evidence that it was written after Toby's death in September 1940. The story takes place in the Twenty-second Century after a Seventh World War, "when nationalism and racism have been outlawed." A highly advanced civilization on Eos (not, however, the Eos of *The Time Stream*) has disseminated throughout the universe a warning of its "fatal mistake" on tablets enclosed in metal "geoids." The discovery of the geoids that reach Earth and the deciphering of the tablets in them is narrated through three characters—Gifford (the head of the deciphering team), his devoted assistant, and a brilliant woman scientist who loves Gifford but kills him because he is incapable of expressing his love for her. Having deciphered the "well-hidden" message of the tablets—"Incapable of saving itself, mere intelligence cannot save others"—she kills herself. (The novel concludes with a "Report of the Asiatic High Command" that thoroughly confounds the reader as to everything that has come before.) *The Cosmic Geoids* is a depressing, despairing story. The "September madness" seems not only general but personal, and the inability to express love is a darker and more explicit theme than in *The Time Stream.*

Although Frederik Pohl found it "an entertaining story [that] will be remembered as one of John Taine's best," Forrest Ackerman found it difficult to believe that John Taine had written it. P. Schuyler Miller lamented, "Time was when each new science fiction novel by 'John Taine' ... could be confidently welcomed as another landmark ... but changes seem to have crept into his more recent work which detract from its lasting qualities." For Richard Witter, *The Cosmic Geoids* "[portended] more and greater things than [John] Taine has ever before attempted (quite a large order, by the way!) but [succeeded] in delivering only frustration and unlimited quantities of confusion."

In 1949 Julius Schwartz contracted with Crawford's company to republish *The Time Stream* and the six novels earlier published by Dutton, that company having turned over its rights to Bell. But only *The Iron Star* and *Green Fire* were published before Crawford was bankrupt. Eshbach's Fantasy Press lasted somewhat longer, but neither Eshbach

---

* The word should have been *Geodes,* but the error passed unnoticed.

nor Crawford could compete when established publishers began to move into the science fiction field.

Williams & Wilkins also returned the rights to *Before the Dawn* as well as to *Queen* and *Handmaiden.*

"The vitality of the books you wrote so many years ago often surprises me," Robert Gill, the president of the firm, wrote to Bell. "I was hellishly disappointed in the reception the world gave *Before the Dawn* when it was first published. I think it was just ahead of its time . . . ."

The publishing and republishing of his novels was satisfying to Bell, but his interest was really with his "main job," which was not the Fermat book but *The Scarlet Night.* He had sent the manuscript to McGraw-Hill in 1949, but it was a year before Edward C. Aswell, the vice president of the firm, responded. Such an exceedingly long poem "posed a number of very unusual problems for a publisher." Although Aswell as well as several members of his staff had read it, they did not feel they could make a judgment, since they were not sure that they had wholly understood it. They had finally contacted the well-known poet Muriel Rukeyser, who was also the biographer of the physicist Josiah Willard Gibbs, and who was well known for integrating contemporary science and technology in her work.

Mrs. Rukeyser was enthusiastic.

"The reading of *The Scarlet Night* has been an extraordinary experience . . . ," she reported to the publisher. "The scene itself reaches the imagination in a curious and piercing way. It is not the scene that Dr. Bell himself thinks it is—Babylon. The scene is dream, and the reverberating images, the sense of time, the evocation of a kind of memory in the possibilities he establishes—all of these are disturbing, terribly disturbing just after a reading, and very likely not to be forgotten.

"At times, the writing, for a short passage, reaches the level of power which these images, and the clashes of meaning, demand. More often, we are given the now-outdated formal balance of [the] period which this poem's date—1910—would indicate. We have, again and again, padded descriptions, lines loaded with modifiers, and always the thumping masculine endings which are hard to take after a few pages—and here they persist for eleven books! But the melodrama of the action, and the tragic drama of the dreams, lifts *The Scarlet Night* above the level of that awkward method, again and again . . . ."

She urged Aswell "to consider future publication of the manuscript in a revised form; to encourage its production as a dramatic work

MURIEL RUKEYSER, known for her interest in science and technology

with music [she suggested an opera]; to explore this direction in publishing, in all of its forms—the meeting place of science and poetry."

Soon she was writing directly to Bell:

". . . it should have been printed in 1910. I suppose they said then that it was fantastic and obscure . . . . Now, of course, it is one of the few poems with any twentieth century thought."

Bell was appreciative of the attention she was devoting to his poem and of her desire to see it published at several levels. What he wanted most, however, was to get the work into print *as it was*. He did not attempt to follow through on any of her suggestions. She, on the other hand, continued—even after Bell's death—to take an active interest in the work, on several occasions reading passages over the radio. Her own subjects at the time—a biography of the anthropologist Franz Boas and a book titled *Dreams and Los Alamos* (Bell helped her to get in contact with Robert Oppenheimer) may have played a role in her response to the poem.

A later reader for Simon and Schuster reported that *The Scarlet Night* was the work of "an earnest and competent but not especially gifted poet," while a reader for the University of California Press noted, "It is a most interesting piece of work, but not really a very good poem."

During Bell's later years his two goals in life were the publication of *The Scarlet Night* and the completion of his book on Fermat's Last Theorem. The latter, which in his prime he would have whipped out in a month, was limping along. One problem was that he wanted to set the mathematical history of Fermat's theorem in the context of the social history of the times. He was having the most difficulty obtaining the latter information about Fermat's own period. Fermat had been a sort of circuit judge. What were the law courts like? How did he travel from place to place? What did he have for dinner?

While he was struggling with these questions, he completed a book that combined *The Queen of the Sciences* and *The Handmaiden of the Sciences* by alternating chapters on pure and applied mathematics. The new title was *Mathematics, Queen and Servant of Science*.

Since the publication of *Queen* in 1931, an event had occurred that had changed mathematicians' most basic conception of their subject. That was Gödel's landmark paper, published the same year, which had shown that in certain respects a system of mathematical logic, such as arithmetic, has an *interior* and an *exterior* (so Bell felicitously put it in his *Development of Mathematics*)—theorems can be exhibited within a system that cannot be proved except by going outside the system.

Since *Handmaiden* had appeared in 1937, there had been another event that was to have an almost equal but quite different significance for mathematics. That was the invention, during the Second World War, of the electronic high-speed computer. At the time Bell was combining *Queen* and *Handmaiden* into one volume, neither he nor any other mathematician—unless it was Alan Turing—had any conception of the role the computer would play in mathematics in the future. Nor did anyone dream that in the space of three decades a famous and hitherto intractable mathematical problem would be "solved" by a computer. In 1951 a computer looked to be simply a machine that could, as Bell wrote, "do in a matter of minutes or hours what no mortal could have achieved in a superhumanly long lifetime."

In 1951, when *Mathematics, Queen and Servant of Science* was making its appearance in bookstores, Bell was concluding his twenty-fifth year at Caltech. He felt that he had never worked for any people "who were more liberal beyond the written word," and he wrote to this effect to President DuBridge.

The following year, on the recommendation of the physicist Robert Bacher, head of the Division of Physics, Mathematics and Astronomy, Bell went on halftime. In 1981, describing himself to Albers as someone "not [having] an enormous rapport with the research [Bell] was doing," Bacher selected as "the most unusual thing about him ... the extraordinarily wide range of things where he knew people and had ideas. They seemed to cover just all sorts of subjects. It's amazing. I don't know that he followed very closely what was happening in the news. I never heard him discuss that subject very much. He usually talked about things where ideas were important in some way. Anything he put his mind to there was a good chance that new ideas would come out of it. That was his nature."

A little earlier Bacher had brought into the mathematical faculty a young number theorist from M.I.T., Tom Apostol, who had been a student of Lehmer's at Berkeley.

"I was to be a kind of replacement for Bell," Apostol explains. "His field was algebraic number theory and mine was analytic number theory, but we hit it off. We had a seminar, just he and I, the first year I was here, on diophantine analysis, where he did most of the talking."

Although Bell was almost seventy at the time of their meeting, Apostol found that the older man had not lost his mathematical skill.

"I remember I was writing a paper and I came across some numbers and talked to him about them. 'Oh yes, those are Stirling numbers

TOM M. APOSTOL
a kind of replacement

of the second kind,' he said. Right away he showed me how they could be generated. He had it right at his fingertips."

The two men soon became friends.

"I would drop over almost every day. I used to take him places. For a drive and things like that. I found him a pleasant curmudgeon. He was a sentimental guy, yes, but he didn't like to show it. He had this tough facade, you know—tough but oh so gentle."

Another view of Bell during his last few years on the faculty comes to me from Mary Pierce, then Mary Kemph. She was Bohnenblust's secretary, and on her own time she had typed *The Scarlet Night* for Bell. One evening, having worked very late, she decided not to drive home but to take "E. T." up on his offer to bed down in his spare room.

"E. T. loved cats, their independent acceptance of humans appealed to him. [So] I drove into the driveway, went to his bedroom window, and meowed until he awoke and said, 'Is that you, Speckled Mouse?' He tottered to the door, let me in, and returned to sleep. He chuckled about what the neighbors would think as I left early in the morning!"

After that, whenever she worked late, she stayed at Bell's.

On occasion, she recalls, he liked to invite all the secretaries to supper. The meal would be prepared by Mrs. Lemons, who would serve drinks, call everybody to the table, and then leave. Bell was "a gracious host," but he also delighted in shocking his guests.

"It all seems so trivial in words," she later wrote in a memoir titled "E. T.," a copy of which she gives me, "but it was a time in my own life when a warm friendship was wonderful. If you could get close to E. T., you were one of the lucky ones!"

In 1953 Bell passed his seventieth birthday. With his retirement imminent his friends discussed a gift. No one is certain who suggested the *Arithmetica* of Diophantus, but Jake Zeitlin managed to obtain a copy of the famous Bachet edition of 1670, which had been issued by Fermat's son just five years after the death of his father. It contained in its margins the notes that Fermat himself had made in the margins of his own copy of an earlier Bachet edition of Diophantus. The book, almost three hundred years old, was passed carefully from hand to hand by Bell's friends for their signatures. These were almost equally divided between male friends and colleagues and the wives and widows of colleagues and friends.

The *Arithmetica* of Diophantus was the most appropriate gift his friends could have chosen; for it was, in all probability, a similar if not so rare edition of Diophantus that he had pored over so intensely and so lovingly more than forty years earlier in the library of Columbia University.

BELL'S MATHEMATICAL PAPERS AS CLASSIFIED BY LINCOLN DURST

Year	Numerical Functions	Forms and Paraphrases	Special Sequences	Diophantine Analysis	Miscellaneous Mathematical Topics	Cumulative Totals
1915	•					•
1916						
1917	•	•				
1918					•	•••
1919	•	•••••			•	•
1920	••	•••			•	•••••••
1921		•••••••	•		•	•••••••
1922	•	••	•••		•	••••••••
1923	•	•••••••••	••			••••••••••
1924	••	••••••••••••	•			••••••••••••••
1925	•	•••	••			•••••••••••••••
1926	•••	••	•••••		••	••••••••••
1927	••••	••				••••••••••
1928	••	••••	•		•••	••••••••••
1929		••••	••••	•	••••	•••••••••••••
1930	••	•••••••	•		••	•••••••••••••
1931	•••	••		•	•••••••	•••••••••••••••••
1932		•••		•	••••	••••••
1933		•	•		••	•••••
1934	•	••••	••	••••	•••	•••••••••
1935		•••••	•			•••••••
1936	•••	•••••				•••••••
1937	••••	••		••		••••••••
1938		••	••	•		•••••••
1939			••	•	•	•••••
1940			••	••		•••
1941			•	••		•••
1942				•		••
1943				•		•
1944			•	•	•••	••••••
1945				•		•
1946				•	••	•••
1947	•	•		•••		•••
1948	•			•••		•••••
1949				••		•••
1950				••		••
1951	•					
1952						•
Total	35	82	29	28	37	211

The Colloquium volume, *Algebraic Arithmetic,* is omitted as not subject to classification because it spans the three main areas of Bell's interest prior to 1927.

The Last Problem

Since Bell was now seventy years old, and emeritus, the Home Secretary of the National Academy of Sciences deemed it appropriate to ask him for personal information that could be used after his death in the Academy's traditional memoir of a deceased member.

But Bell was not to be trapped.

"You will recall that the late H. S. White wrote his own biographical sketch," he reminded the Secretary. "I plan to do the same thing, as none of my friends have very much relevant data. Does the Academy want a complete list of publications, or will a short selection do?"

The Secretary assured him that the Academy would be most pleased to receive an account such as the one he proposed and that it would like the bibliography to be as complete as possible, since it would constitute a permanent record of his writings. About a year later, when neither the promised account nor the bibliography had been received, the Secretary wrote again to remind Bell. He received no answer.

The National Academy's file on Bell contains two items of personal information that he furnished at earlier times. One is a curriculum vitae submitted in 1927 when he was elected to membership. It is factually accurate except for the statement that he attended the University of London from 1900 to 1902. The other is a questionnaire compiled for the Academy by the Eugenics Society, to which he responded.

Receiving a copy of this questionnaire ("Abridged Record of Family Traits") four years into my search for Bell, I am quite excited. Up to this point, except for the quoted notes in *Twentieth Century Authors*, I have seen nothing that Bell actually put into writing about his family

background. The questionnaire is rather unscientific. I am surprised
that he did not toss it away, but for some reason he did not. Although
he explained that his personal records had been lost in the San Francisco earthquake and fire and that he had not kept track of his family—
"nor they of me"—he nevertheless made a stab at completing all four
pages. These asked for data about his parents, himself and his wife,
and his offspring as well as his paternal and maternal grandparents.

He answered questions about his paternal grandfather, James Bell
Sr., in some detail. *Occupations at successive ages* are given as "Merchant,
Shipping, Politics, Diplomacy," and *Special tastes, gifts or peculiarities of
mind or body* as "Literature, great business ability." Not unexpectedly,
there is no mention of the fishing industry as the source of the family's
income; but the addition of "Diplomacy" to James Bell Sr.'s *Occupations
at successive ages* is unexpected. In a lengthy description of the funeral
of "Mr. Deputy Bell" that appeared in the *Norwood News,* there is nothing to indicate that he ever practiced diplomacy other than as a member of the Fish Trades Association and the Central Hill Baptist Church,
where he was a deacon. There are a number of slightly surprising in-
accuracies in Bell's answers in regard to his paternal grandmother. He
gave her maiden name as *Kent* rather than *Redward*—odd, since Redward was his older brother's middle name. He also listed her as having
had five boys and three girls when in fact James Bell Jr., Bell's father,
was one of two brothers and half a dozen sisters. It appears that he
did not get to know his many Bell aunts in Upper Norwood when he
returned to England after his father's death.

In the information he furnished about the maternal side of his
family, he mentioned nothing about Peterhead. According to him, everyone in the family was born in Aberdeen and continued to reside
there. He described his mother's mother (who died before his own
parents were married) as "Clear writer (as shown by letters), good lin-
guist" and his mother as "Very clear writer (as shown by letters, etc.)."
It was only his mother's father, James Lyall, the parish schoolmaster,
about whom he wrote specifically, although he had never known him
either. He had been a "Greek and Latin teacher" and "a distinguished
classicist [who] also trained several good mathematicians (3 were Cambridge wranglers, 2 'Seniors')."

I turn most eagerly to Bell's answers to questions about his father.
They reveal that he was aware of his father's first marriage and of the
daughter who died as an infant—in short, he knew the source of the
name Temple when he took it as his nom de plume. But such things are

not what I am after. Perhaps here at last, in this questionnaire that he filled out forty-five years after his father's death, I shall learn the cause of that death and, as a consequence, the reason that Bell memorialized it so frequently and so cryptically.

In the portion of the questionnaire pertaining to his father, Bell answered every question. Beginning with *Year of Birth* and *Birthplace*, both accurate, I hurry on. *Cause and age of death* "Angina pectoris, 59y. 11mo." The age is precise to the month. Do I actually have in hand the answer I have sought for so long as to the cause of the death of James Bell Jr. on January 4, 1896, in San Jose, California?

In view of Bell's answers to the other questions about his father, I am not at all sure. It is quite likely that James Bell Jr. suffered from angina pectoris, as apparently did Bell himself. In fact, he referred to his "heart attack" of 1932 parenthetically as "angina." There is a line in *Desmond* where he writes that the villain Mudge, hearing Desmond's father cry out at the news of his son's supposed death, "had [once] heard a man in the grip of angina pectoris, and the cry that had awakened him had seemed strangely like that other." If James Bell Jr. did die a natural death, it is quite possible—since he died without a will—that he died of a coronary thrombosis, which although medically identified by that time was not commonly diagnosed. Yet I can't rid myself of the nagging fact that Dr. R. E. Pierce, passing the County Records Office on his way to the County Courthouse, did not stop to record the death and that Helen Bell paid out *a thousand dollars* to return her husband's body to England to lie beside his first wife and their infant daughter.

Bell's answers to the other questions about his father do not inspire confidence:

Education *For navy*
Favorite studies or pursuits *Shipping, business, minor diplomacy*
Tastes, gifts or peculiarities of mind or body *Good navigator (first class), good mathematician, good in business*

It is indeed a curious document, which Bell not only completed but also returned to the National Academy of Sciences for its archives. The answers, where I am familiar with the facts, are not outright lies. He does not *write* that his father was "the owner of a fleet of sailing vessels," although he *told* many people that he was. Rather he gives very general answers to the question regarding *Occupations at successive ages*; but by twisting or omitting specific facts, he implies that his father was something different from the actuality: Peterhead fishcurer, San Jose orchardist.

After examining this document, from which I have hoped to gain so much information, I have an opportunity to talk again to Professor and Mrs. Dilworth. I ask if they ever felt that Bell was being secretive about his youth.

"Well, I think, if I were to judge, that his youth had not been terribly happy," Dilworth replies.

Miriam Dilworth agrees.

"Did you ever ask him about his youth?"

"You would never *ask*," she says. "I don't know why. You just felt that you shouldn't. You sensed it very quickly."

I then ask Dilworth, directly, if he ever had any reason to suspect Bell of lying.

"There was an occasion or two where I think he stretched the truth a little bit, but not really lying. ["Not an out-and-out lie," inserts Miriam Dilworth.] He wouldn't stoop to subterfuge or something like that. He was too open and honest."

This is, in essence, the same report I receive from others of whom I ask the same question.

On June 14, 1953, only a week after Bell received the *Arithmetica* of Diophantus from his friends, his longtime colleague, A. D. Michal, died. He had been ailing for sometime from heart trouble, and many people blamed the tensions in the mathematics faculty during the past ten years for his death at fifty-four.

Michal had been a prolific researcher and author as well as adviser to twenty-eight Ph.D.'s. These recall his encouragement and his generosity with appreciation in spite of what some see as his weaknesses. He gave young people the idea that research was something they could do themselves, according to D. H. Hyers, who wrote several papers with him. He also had "an intense devotion" to mathematics— "it was something very personal with him, a part of his life, like a religion," Angus Taylor recalled at the memorial meeting of the "Peripatetic Seminar," a group organized by Michal that met successively at Caltech, UCLA, and USC—and called themselves, in a play on his name, "followers of Aristotle."

"Michal was a person of great ability, which is probably what Bell saw in him when he brought him to Caltech," another student of Michal's, Edmund Pinney, contributes, "but he also had a flaw which

THE HUMBLE SOURCE
of number theory

was very detrimental to him as a mathematician: a kind of childish desire for glory."

Michal's widow, Luddye, had been a support to Bell after Toby's death—"the only person he would see"—and he supported her in her loss. Almost immediately it became their custom, for as long as he continued to live in Pasadena, to dine together on Saturday or Sunday, his housekeeper, Mrs. Lemons, having left behind a dinner that needed only to be heated. Frequently on these occasions Bell asked Mrs. Michal to help him organize materials about his family.

"He had me going through some files, but it was always going through the same files. He told me some things about his family. He came to this country, you know, never to return, never wanted to."

I ask her if he ever told her how he happened to come to Stanford.

"He did say that he had letters of recommendation from his parents to some of their friends in America. They may have suggested Stanford. I think they had a big fruitpacking company."

I suggest Del Monte, and she says no, it was an English company.

When she came to dinner, she recalls, she usually stopped on the way to buy a bottle of Jack Daniel's at his request.

"I tried to water his drinks," she laughs, "but he always knew."

Mrs. Lemons, whom Cagey had hired to work for two weeks after the mugging in Redondo Beach, remained for fourteen years to

E. T. BELL with his longtime housekeeper, Mrs. Frances Lemons

cook and care for Bell's house, to break most of his dishes, and to keep his checkbook in balance. (Contrary to popular opinion, mathematicians are sometimes poor in matters of everyday arithmetic.) On occasion she also shared a drink with him. There was always some excuse. Toby's birthday. The date of their anniversary. The date of her death.

"I think it was the bourbon that he drank almost continuously in later years that kept him going," Apostol tells me. "He wasn't a drunkard in the usual sense, reeling and staggering and slurring his words. He could always keep the drinking under control, and he had his wits about him. He was very sharp even then. But I think he just lived on bourbon."

Smiling at the memory, he recalls an occasion when Bell, glancing down at the liquid in the glass in his hand, muttered, "Damn! I forgot to put any water in this." He went to the kitchen and in a moment returned with the glass full of an even darker liquid.

He was deeply hurt when the Institute's administration decided that, since he lived just across the street, he no longer needed an office. In retrospect, taking Bell's office from him has been recognized as an insensitive act, and since that time every professor at Caltech upon becoming emeritus has retained an office if he wishes one.

In addition to "organizing" his family papers (most of the pencilled notes in books and on pictures that I have seen at Watsonville were probably made at this time), Bell began to clean out his house, One day when Apostol dropped by, he announced that he had had "a bonfire" and burned all his correspondence.[*]

"*What!*" exclaimed Albers, who was told this by Apostol during their 1981 interview.

"Yes. He just wanted to get rid of it, he said. He must have had fabulous letters from people like Landau and Hardy. He wanted to burn his books too—although I don't think he was serious about that—but I said, 'You don't want to do that.' 'Well, I don't know what to do with them.' I said he should give them to the mathematics faculty. I was the library representative for mathematics at that time so I got the idea, instead of sticking these books in the library as duplicates, why didn't we start a little research library of our own. A little library that didn't circulate except within our building. I hauled his books over, I remember. But he saved some for himself. He saved the Diophantus—

---

[*] Backyard incinerators were commonly used at that time to dispose of garbage and wastepaper.

NUMBER THEORY CONFERENCE AT CALTECH IN 1955

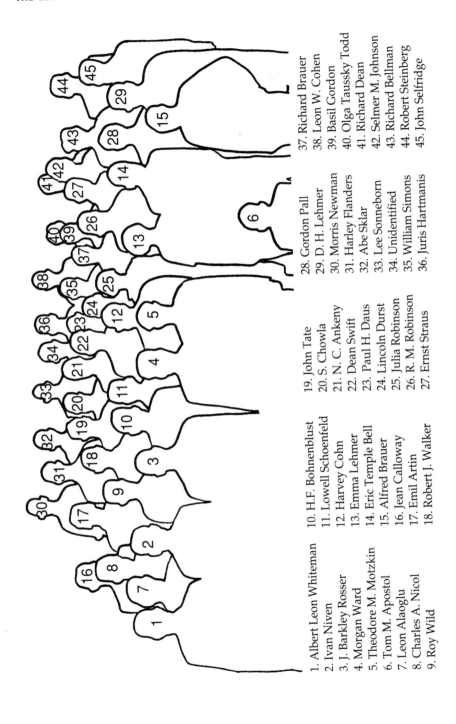

1. Albert Leon Whiteman
2. Ivan Niven
3. J. Barkley Rosser
4. Morgan Ward
5. Theodore M. Motzkin
6. Tom M. Apostol
7. Leon Alaoglu
8. Charles A. Nicol
9. Roy Wild
10. H.F. Bohnenblust
11. Lowell Schoenfeld
12. Harvey Cohn
13. Emma Lehmer
14. Eric Temple Bell
15. Alfred Brauer
16. Jean Calloway
17. Emil Artin
18. Robert J. Walker

19. John Tate
20. S. Chowla
21. N. C. Ankeny
22. Dean Swift
23. Paul H. Daus
24. Lincoln Durst
25. Julia Robinson
26. R. M. Robinson
27. Ernst Straus
28. Gordon Pall
29. D. H. Lehmer
30. Morris Newman
31. Harley Flanders
32. Abe Sklar
33. Lee Sonneborn
34. Unidentified
35. William Simons
36. Juris Hartmanis

37. Richard Brauer
38. Leon W. Cohen
39. Basil Gordon
40. Olga Taussky Todd
41. Richard Dean
42. Selmer M. Johnson
43. Richard Bellman
44. Robert Steinberg
45. John Selfridge

that pleased us, the fact that he kept the book we had given him and gave it to his son."

After Bell's retirement the letters to Taine and Janet became briefer and less frequent. The young people tried to keep track of him by telephone, for he was frequently unwell. He became very thin, probably as a result of his drinking, and "tottery." He continued to work doggedly on *The Last Problem*.

When Thoralf Skolem, who had done considerable work on undecidability, visited Pasadena, Bell asked him whether he thought it was possible that Fermat's Last Theorem might be "one of Gödel's undecidables."

"He got quite excited and said, 'It might well be!'" Bell reported to H. S. Vandiver, who had done considerable work on the theorem itself. "Now if so, where does that leave F. in his famous assertion [that he had proved the theorem]?"*

There continued to be publishing and republishing of Bell's earlier books, both the popularizations and the novels. His science fiction story *G.O.G. 666* ("General Order of Genetics" and "the number of the beast") appeared in 1954. Although it was his last novel, and reviews were uniformly unfavorable, there continued to be notice of John Taine's work in science fiction magazines as reprints of earlier novels appeared. He was "the old master," respected as a pioneer in the field, recalled frequently with a "vague but steadfast affection." But his books were not much read. His writing was described as "prolix," his characters as "stick figures." The latter was not quite fair, for many characters remain in one's mind long after the plots have been forgotten. Postwar readers were offended by his treatment of people of other races. There was no question but that John Taine, who had never dealt with hardware in his stories, was out-of-date when in October 1957 the Soviet Union launched the first artificial satellite, *Sputnik*.

One morning the following year Mrs. Lemons, coming to work, found Bell unconscious on the kitchen floor. He had fallen, having climbed up on a chair to change a light bulb, and had broken his shoulder. Taine and Janet came down to Pasadena. Over the protests of Bob and Angela Robertson and others who thought he should remain at

---

* In fact, if Fermat's Last Theorem were undecidable in some formal system of arithmetic, it would have to be true, since a solution to the equation could be exhibited if it were false.

home where he would have his friends about him, they took him back with them to Watsonville.

By the summer of 1959 he was in sufficiently good health to return to Pasadena, but he was not really well. Apostol, who stopped at his house almost every day, commented upon the fact that he was no longer shaving. Was he raising a beard? No, he just always cut himself when he shaved because his hand shook.

"So I brought over my electric shaver, but it was clear he didn't have enough control to shave even with an electric shaver. So then I would shave him."

Soon it was necessary for Taine and Janet to take him back again, permanently this time, to Watsonville. They arranged for him to have a room in the hospital where they both worked and surrounded him with his books—the collected reprints, the novels, the popular books on science—although what he read was poetry. Sometimes he hosted a kitten on his bed. He had his bourbon and his cigars, but after he set the bed on fire Janet laid down the law: no smoking unless someone else was in the room.

Durst, travelling back to Southern California from Palo Alto, stopped to pay a visit, but Bell was asleep and he did not waken him. He still has a picture in his mind of the old sleeping Bell in the hospital room, which he seemed "to fill to the brim."

At that time Janet, with three small children, devoted only the mornings to her pediatric practice. She regularly lunched at the hospital with her father-in-law and then took him for a drive. Watsonville was then an even more rural community than it is today, its many orchards spread over a California valley very similar to the nearby Santa Clara Valley. Bell enjoyed the drives, but nothing he ever said gave Janet the slightest hint that he had spent twelve years in such country when he was a boy.

At the beginning of his stay at the hospital she had brought one or more of the children to visit him from time to time. But he was no better a grandfather than he had been a father. After he cut off little Lyle, who was telling him about something that had happened at school, with an abrupt "I'm not interested in that!" she ceased the custom.

Even in the hospital he continued to work on *The Last Problem.* By late 1960 he had almost finished the book, but it was clear that he would not be able to write a final chapter. Taine arranged with D. H. Lehmer to prepare some material that would bring the work on Fermat's theorem up-to-date and bring the book to a conclusion. On December 5, sitting

up in his hospital bed, Bell signed a contract with Simon and Schuster
for the publication of *The Last Problem*. Fifteen days later on December
20, 1960, seven weeks before his seventy-seventh birthday, he died.

Once, some years after Toby's death, walking home with Lee
DuBridge after a memorial service for their friend Richard Chace
Tolman, Bell had remarked that it was rather nice to have a kind of
punctuation at the end of a life. But he himself had remained firm as
to his own final wishes. No funeral, no memorial service, cremation of
his remains, the ashes to be scattered with Toby's outside Yreka.

His final book, *The Last Problem*, was published the year following
his death. Thirty years later it was reissued by the Mathematical Association of America with an introduction by Underwood Dudley—who
had some difficulty in describing it.

"It is not a book of mathematics. Pages go by without an equation
appearing, and in mathematics books you are not told such things as
that the ancient Spartans were 'as virile as gorillas and as hard (including their heads) as bricks.' ... It is not a history book, either. History
books do not contain the nine equations in ten unknowns that come
from the cattle problem of Archimedes. ... It is not a history of number
theory because it includes too much about the history of the western

D. H. LEHMER
wrote final chapter

world, and it is not a history of western civilization because its focus is on mathematics. It is too entertaining to be scholarly and contains too much mathematics to be widely popular.

"It is an unusual book," Dudley concluded—as unusual as the man who had written it.

For some ten years Bell's ashes remained behind a hatbox on a closet shelf in Watsonville. Finally Taine and Janet, two busy doctors with a growing family, had time to drive up to Yreka. Following Bell's 1941 map and photos he had taken on a 1948 trip to Seattle, they located the spot where he had scattered Toby's ashes. But the town had grown, the area was no longer isolated, children had built a treehouse in what had once been a sapling. They parked their yellow Oldsmobile and, feeling very conspicuous, ducked under an electric fence and ran up the hill to the specified stand of rock, leaving behind a white trail of calcium pellets and bits of bone. They could not help thinking how Romps would have crowed with delight at their discomfort.

Now, more than thirty years after Bell's death, it appears that a memoir of his life and scientific work will finally be written for the National Academy of Sciences, correcting the false impression given by the inadequate and inaccurate summary in the *Dictionary of Scientific Biography*. A new look will have to be taken at the science fiction of John Taine as a result of discoveries about the dates of the writing of his novels, particularly *The Time Stream*; for these indicate that he was more of a pioneer in that field than has been thought.

As I write, I am looking back on almost five years devoted to a search for the man who was E. T. Bell. When I drove down to Watsonville in the spring of 1988, I had no intention of doing anything more than a profile of a somewhat eccentric mathematician who wrote popular books on his subject and science fiction novels. Yet somehow, after I took that first step, in spite of my original intention, I was not able to cease in my search for the man who had hidden a dozen significant years of his life from his beloved wife and his only son. A life cannot be left half finished, so I was to find out; yet I cannot help thinking that it might have been quite possible for me to have written a profile of Bell that would have repeated much of the misinformation already in print about him:

If Lyle Bell had not gone to Bedford to find the grave of his great-grandmother—,

If Helen Bell (or had it been Enid?) had not had placed on her grave the stone with the identifying lines "widow of James Bell of San Jose, California"—.

Yet even with the knowledge that Bell's father had emigrated to San Jose, I needed to have a date for his death there before I could proceed. That Bell himself had unwittingly provided in the lightly penciled "In Mem." notes on the title page of his father's Caesar. Even then, with the knowledge that Bell's father had emigrated to San Jose and that the "Mrs. James Bell" who had returned to England after his death was the mother of E. T. Bell, I would never have suspected that Bell and his siblings grew up in San Jose if Bell himself had not given the game away once again: first, by detailing his early education so specifically and unnecessarily as to years "(all)" for the administration at Caltech, and then by describing straightforwardly, but also unnecessarily, his first contact with Edward Mann Langley at the Bedford Modern School "as a new boy in the Upper Fifth Form."

It was "a tangled web" Bell had woven, and it was the untangling that had challenged me to take a greater interest in his life and to devote more effort to researching it than I had ever intended. I had set out to catch a cat that could not be caught, an absorbing kind of game, and had ended up with a gifted, complex, sensitive, and very vulnerable human being in my net—a man who offended many and yet who was so loved by others that they find it difficult to convey the depth of their feeling for him even more than thirty years after his death.

Why had Bell so early and so long practiced to deceive?

It is a question I have to admit I cannot answer.

Yet even a seemingly trivial mystery like how he happened to lose the first joint of his right thumb nags at me. The many different stories he told about it seem so pointless, even as "spontaneous fictions"— unless there was a point. The most frequently told story—repeated in print in the Caltech Alumni Association's magazine—is that he lost the thumb while working at a job in a lumbermill in Siskiyou County. But such employment is not mentioned among the miscellaneous jobs he lists in the autobiographical notes for *Twentieth Century Authors*. It is listed only in biographical entries where the information came either orally from Bell, or clearly from Taine. The earliest photograph of him, at eighteen, shows him holding his pipe in his left hand, his right hand hidden. Did he lose his thumb in the "hard time" of his youth, as Miriam Dilworth thinks? Oddly enough, he did not mention the missing thumb when asked about identifying marks at the time of his naturalization in 1920. Did the loss have something to do with the death of his father? An accident? I can't say, and yet I feel that it is, somehow, a clue.

One afternoon, checking the responses of readers to *The Time Stream* in Charles Brown's copies of *Wonder Stories*, I mention to Brown, who is editor of the science fiction newsletter *Locus*, that "John Taine" had in essence "made up his life."

Brown chuckles.

"Almost every science fiction writer I know makes up the story of his life," he says. "It is as if he has to make a story out of it. To bring past, present, and future into a unified whole."

"But this science fiction writer was a mathematician," I object and, echoing Tony Rothman's comment on Bell's fictionalization of the life of Galois, "a member of a profession dedicated to Truth. A mathematician does not customarily play fast and loose with facts."

Brown concedes that not many of the science fiction writers he knows are mathematicians.

The story that Bell concocted about his early life—the English-educated young man emigrating to America to avoid going into his father's shipping business—does not seem to me so interesting as the story I have finished recounting, and I do not believe that he made it up for literary reasons.

"It is a serious and humiliating thing," says the younger brother in *The Forbidden Garden*, "to reveal family secrets."

Were there secrets in Bell's family life that he considered humiliating? I have found out a great deal about his background—his grandfathers, his father and mother, his sister and brother, even his father's first wife—and in nothing that I have found (from my point of view or that of his own family) is there anything for him to be ashamed of. One thing that is quite clear is that he was embarrassed by his family's connection with the fishing industry, successful as it was. Among his memorabilia there is what might be called a leather "secretary" to hold a pen and an account book. Inscribed on it in gold are the words "J. Bell, Dep., 1 Billingsgate." The gold letters that form *Billingsgate* have been scraped away with something like a knife, but there is enough left for it to be deciphered by anyone who knows that E. T. Bell's grandfather was a fish factor at the Billingsgate Market in the nineteenth century—"Mr. Deputy Bell."

Since my initial visit to the Bells' home, I have come a number of times—to report on progress, to ask questions, to borrow first editions,

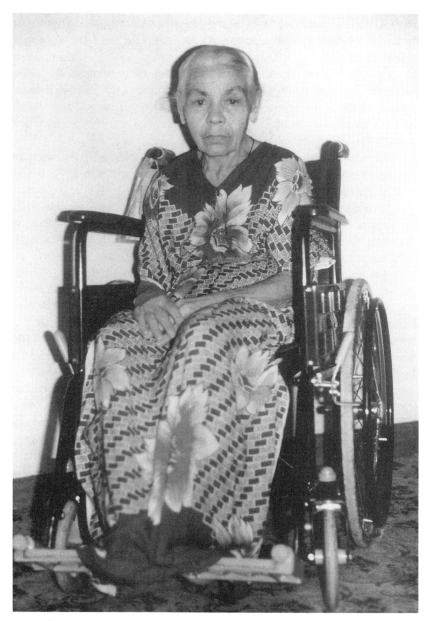

BEATRICE BELL ROBERTS, only surviving child of James Redward Bell

to pore over memorabilia, to test hypotheses. Much has happened there since 1988. Taine has suffered a heart attack and a serious stroke from which, happily, he has made a quite good recovery. Janet has retired from her medical practice. The house has been severely shaken in the La Prieta earthquake of 1989—Bell's books as well as others thrown from their shelves, some of his models of dinosaurs and his large oriental bowls broken, although since repaired. Taine and Janet now have a granddaughter, their first, to whom Lyle and his wife (unaware of Bell's "spontaneous fiction" about his mother's name) have given the name Heather.

Thanks to letters from Jill Morgan and her sister, Vivien Drayson, I have been able to tell Taine something about the life of his father's sister, Enid. Then, following the discovery in Sri Lanka of descendants of Bell's older brother, James Redward Bell, I have brought him pictures of the brother and friendly letters from W. A. Roberts, writing for his wife, Beatrice, the only surviving child of James Redward Bell. Taine and Janet are welcoming to their newly found relatives. Janet makes copies of family pictures and, drawing up a Bell family genealogy based on my research, sends them to Sri Lanka. With the cooperation of Arthur C. Clarke, who has played the decisive role in locating James Redward Bell's family, we arrange for the purchase of a wheelchair for the daughter, Beatrice, who is so badly crippled with arthritis that she has been confined to her bed. Her husband writes gratefully, "It has brought new meaning to her life and joy to our family." And so, some thirty years after Bell's death in Watsonville, the family that he tried so stubbornly to keep apart has been brought together.

As for Bell himself—he has gone to his death with his secret, whatever shame or hurt or joy it was that he carried away from a California valley when he was thirteen years old.

INDEX

Note: Photographs and other illustrations are in boldface type.